ESAs MADE EASY

A Checklist Approach to Phase I Environmental Site Assessments

André R. Cooper, Sr., R.E.A.

Government Institutes
Lanham • Boulder • New York • Toronto • Plymouth, UK

Published in the United States of America
by Government Institutes, an imprint of The Scarecrow Press, Inc.
A wholly owned subsidiary of
The Rowman & Littlefield Publishing Group, Inc.
4501 Forbes Boulevard, Suite 200
Lanham, Maryland 20706
http://www.govinstpress.com/

10 Thornbury Road
Plymouth PL6 7PP
United Kingdom

British Library Cataloguing in Publication Information Available

Library of Congress Cataloging-in-Publication Data

978-0-86587-536-4

∞™ The paper used in this publication meets the minimum requirements of American National Standard for Information Sciences—Permanence of Paper for Printed Library Materials, ANSI/NISO Z39.48-1992.
Manufactured in the United States of America.

General Contents

Table of Contents

About the Author

André R. Cooper, Sr. is the president and CEO of *Cooper's Information and Research Services, L.L.C. (Coopers)* and *LTR Environmental, Inc. (LTRE)*, a Denver-based environmental consulting firm specializing in resolving compliance issues for government agencies, real estate lenders, asset management firms, and institutional and private property owners nationwide. Mr. Cooper is responsible for client liaisons and for managing a multidisciplinary project management/information research team. He is an environmental compliance specialist with experience in implementing complex *multi-media* environmental projects, nationwide.

Mr. Cooper has a Master's degree from the College of Architecture and Urban Planning at the University of Washington, Seattle; and a Bachelor's degree from the Department of Environmental Planning and Management at the University of California, Davis. He is a Registered Environmental Assessor and a member of the National Association of Environmental Professionals.

Mr. Cooper has prepared numerous environmental deliverables for the public and private sectors, and is the author of: **Cooper's Pocket Environmental Reference, Cooper's Pocket Environmental Compliance Dictionary, Cooper's Comprehensive Environmental Desk Reference w/Supplemental Spell Check Disk, Cooper's Comprehensive Environmental Desk Reference on CD-ROM**, and **Cooper's Acute Toxic Exposures Desk Reference w/CD-ROM.**

Coopers/LTRE is available to assist financial institutions, novice and experienced environmental personnel, and real estate professionals nationwide with Phase I ESA compliance and preparation. *Coopers/LTRE's* **VP ESA Service** is an automated environmental research and study preparation service in which project plans, oversight and field checklists are provided. Database searches and USGS maps are also part of the **VP ESA Service**. When a client returns completed project plans, photos and checklists, to *Coopers/LTRE,* we analyze the findings, compile a report and submit an 'electronic' Phase I ESA study.

Now, appraisers and novices can conduct field tasks for an ESA and have their information analyzed by professionals with years of experience! This can effectively reduce the cost of an ESA by up to 40%. For additional information contact *Coopers/LTRE* by FAX at (303) 364-2834.

Preface

In 1989 EPA's Office of Solid Waste and Emergency Response (OSWER) issued *Directive No. 9835.9*, which provided *Guidance on Landowner Liability* under Section 107(a)(1) of CERCLA, *De Minimis Settlements* under Section 122(g)(1)(B) of CERCLA, and *Settlements with Prospective Purchasers of Contaminated Properties.*

This guidance set forth EPA's policy on issues of landowner liability and settlement with *de minimis* landowners under CERCLA, and provided a brief discussion and policy statement concerning settlement with prospective purchasers of contaminated property. This guidance also analyzed the language in CERCLA Sections 107(b)(3) and 101(35), which provide landowners certain defenses to CERCLA liability.

The purpose of the directive was to provide general guidance on landowner liability under the Comprehensive Environmental Response, Compensation, and Liability Act of 1980 (CERCLA), as amended by the Superfund Amendments and Reauthorization Act of 1986, Pub. L. No. 99-499 (SARA), 42 U.S.C. §9601 *et seq.,* and to provide specific guidance on which landowners qualify for *de minimis* settlements under Section 122(g)(1)(B). This marked the beginning of what is now known as the *Innocent Landowner Defense* and prompted the need for Phase I Environmental Site Assessments (ESAs).

The brief excerpt from CERCLA Section 107 following this preface exemplifies why this book, the second in Government Institutes' Made Easy series is needed: busy compliance professionals and personnel who are new to environmental regulations need an easy to follow, step-by-step guide to understanding their responsibilities under these complex regulations.

The checklist approach to conducting Phase I Environmental Site Assessment presented in this book will allow both novice and experienced environmental and real estate professionals to simplify the ESA preparation process. It also explains in detail what constitutes a high-quality ESA, making this an indispensable aid to anyone who evaluates ESAs.

ESAs Made Easy is being sold with the understanding that neither the publisher nor the author is engaged in rendering legal advice or services of any kind, and no endorsement by EPA, or any other agency, on the information herein, implicit or explicit, has been received.

André R. Cooper, Sr., R.E.A.

COMPREHENSIVE ENVIRONMENTAL RESPONSE and LIABILITY ACT

Excerpt from 42 U.S.C. 9607 [CERCLA §107]

Sec. 9607. Liability

(a) Covered persons; scope; recoverable costs and damages; interest rate; "comparable maturity" date.

Notwithstanding any other provision or rule of law, and subject only to the defenses set forth in subsection (b) of this section—

(1) the owner and operator of a vessel or a facility,

(2) any person who at the time of disposal of any hazardous substance owned or operated any facility at which such hazardous substances were disposed of,

(3) any person who by contract, agreement, or otherwise arranged for disposal or treatment, or arranged with a transporter for transport for disposal or treatment, of hazardous substances owned or possessed by such person, by any other party or entity, at any facility or incineration vessel owned or operated by another party or entity and containing such hazardous substances, and

(4) any person who accepts or accepted any hazardous substances for transport to disposal or treatment facilities, incineration vessels or sites selected by such person, from which there is a release, or a threatened release which causes the incurrence of response costs, of a hazardous substance, shall be liable for—

(A) all costs of removal or remedial action incurred by the United States Government or a State or an Indian tribe not inconsistent with the national contingency plan;

(B) any other necessary costs of response incurred by any other person consistent with the national contingency plan;

(C) damages for injury to, destruction of, or loss of natural resources, including the reasonable costs of assessing such injury, destruction, or loss resulting from such a release; and

(D) the costs of any health assessment or health effects study carried out under section 9604(i) of this title.

The amounts recoverable in an action under this section shall include interest on the amounts recoverable under subparagraphs (A) through (D). Such interest shall accrue from the later of (i) the date payment of a specified amount is demanded in writing, or (ii) the date of the expenditure concerned. The rate of interest on the outstanding unpaid balance of the amounts recoverable under this section shall be the same rate as is specified for interest on investments of the Hazardous Substance Superfund established under subchapter A of chapter 98 of title 26. For purposes of applying such amendments to interest under this subsection, the term 'comparable maturity' shall be determined with reference to the date on which interest accruing under this subsection commences.

(b) Defenses

There shall be no liability under subsection (a) of this section for a person otherwise liable who can establish by a preponderance of the evidence that the release or threat of release of a hazardous substance and the damages resulting therefrom were caused solely by—

(1) an act of God;

(2) an act of war;

(3) an act or omission of a third party other than an employee or agent of the defendant, or than one whose act or omission occurs in connection with a contractual relationship, existing directly or indirectly, with the defendant (except where the sole contractual arrangement arises from a published tariff and acceptance for carriage by a common carrier by rail), if the defendant establishes by a preponderance of the evidence that (a) he exercised due care with respect to the hazardous substance concerned, taking into consideration the characteristics of such hazardous substance, in light of all relevant facts and circumstances, and (b) he took precautions against foreseeable acts or omissions of any such third party and the consequences that could foreseeably result from such acts or omissions; or

(4) any combination of the foregoing paragraphs.

ESAs MADE EASY

A Checklist Approach to Phase I Environmental Site Assessments

Section 1

Introduction to ESAs

The purpose of conducting an ESA is to: 1) identify existing or potential environmental hazards; 2) identify resources with natural, cultural, recreational, or scientific values of special significance (i.e., "special resources" for a subject property); and 3) recommend whether further investigation is required.

1.1 CERCLA Overview

- ☐ The Comprehensive Environmental Response, Compensation, and Liability Act (CERCLA) states that in the event of a release or threatened release of a hazardous substance, owners of property where such substance(s) has been "deposited, stored, disposed of, or placed, or otherwise come to be located" are strictly liable for the costs of response.

- ☐ Under CERCLA Section 107(b)(3), the liability generally extends to releases which are caused by a third party "in connection with a contractual relationship, existing directly or indirectly" with the owner.

- ☐ Congress, in the Superfund Amendments and Reauthorization Act (SARA), clarified the defense to liability available to landowners under Section 107(b)(3) by specifically defining the term "contractual relationship" to include deeds and other instruments transferring title or possession unless the landowner can demonstrate that at the time he/she acquired the property, he/she had no knowledge or reason to know of the disposal of the hazardous substances at the facility.

- ☐ Based upon SARA, any person who acquires already contaminated property and who can satisfy the remaining requirements of CERCLA Section 101(35) as well as those of Section 107(b)(3) may be able to establish a defense to liability.

1.2 Landowner Liability

- ☐ Section 101(35)(A) of CERCLA, as amended by SARA, states that a real estate deed represents a contractual relationship and specifically defines "contractual relationship" to include "land contracts, deeds, or other instruments transferring title or possession" (for example, leases) unless the property was acquired after the disposal or placement of the hazardous substance which is the subject of the release or threat of release *and* the landowner establishes by a preponderance of the evidence that:

  - at the time the defendant acquired the facility, the defendant did not know and had no reason to know that any hazardous substance, which is the subject of the release or threatened release, was disposed of on, in, or at the facility;
  - the defendant is a government entity which acquired the facility by escheat, or through any other involuntary transfer or acquisition, or through the exercise of eminent domain authority by purchase or condemnation; or
  - the defendant acquired the facility by inheritance or bequest

- ☐ In addition, the landowner must satisfy the *due care requirements* of Section 107(b)(3) in order to establish the third party defense. The third party defense does not apply if the party purchased the property "with actual or constructive knowledge that the property was used for the generation, transportation, or storage of any hazardous substance."

1.2.1 Due Care Requirement

- ☐ The requirements which must be satisfied in order for the Environmental Protection Agency (EPA) to consider a settlement with landowners under the *de minimis* settlement provisions of Section 122(g)(1)(B) are substantially the same as the elements which must be proved at trial in order for a landowner to establish a third party defense under Section 107(b)(3) and Section 101(35).

- ☐ Before the Agency will approve settlements with owners of contaminated property, several questions concerning landowner eligibility for settlements must be answered, bearing in mind that Section 122(g)(1)(B) does not extend to any party who contributed to the release or threat of release "through any act or omission."

1.2.2 Due Care Demonstration

- ☐ The primary question is: Did the landowner acquire the property without knowledge or reason to know of the disposal of hazardous substances? Section 122(g)(1)(B) applies only to owners who purchased the property without "actual or constructive knowledge that the property was used for the generation, transportation, storage, treatment, or disposal of any hazardous substance."

- ☐ Section 101(35) expressly provides that in order for a defendant to prove that he/she had *no reason to know* of the disposal of hazardous substances, *he or she must demonstrate by a preponderance of the evidence that, prior to acquisition, he or she conducted all appropriate inquiry into the previous ownership and uses of the property consistent with good commercial or customary practice.*

1.2.3 Appropriate Inquiry

- ☐ Under Section 101(35)(B), the obviousness of the presence or likely presence of contamination at the property, and the ability to detect such contamination by appropriate inspection, were key issues. This clearly indicates that a determination as to what constitutes *all appropriate inquiry* under all the circumstances has to be made on a property-by-property site assessment.

- ☐ The Agency typically requires a more comprehensive inquiry for those involved in commercial transactions than for those involved in residential transactions for personal use. The determination was predicated on the basis of what was reasonable considering all of the circumstances.

1.3 Third Party Defense

- ☐ Lenders are also eligible for *third party defenses* and *de minimis settlements* in some circumstances. A lender who does not participate in the management of a facility and who only holds

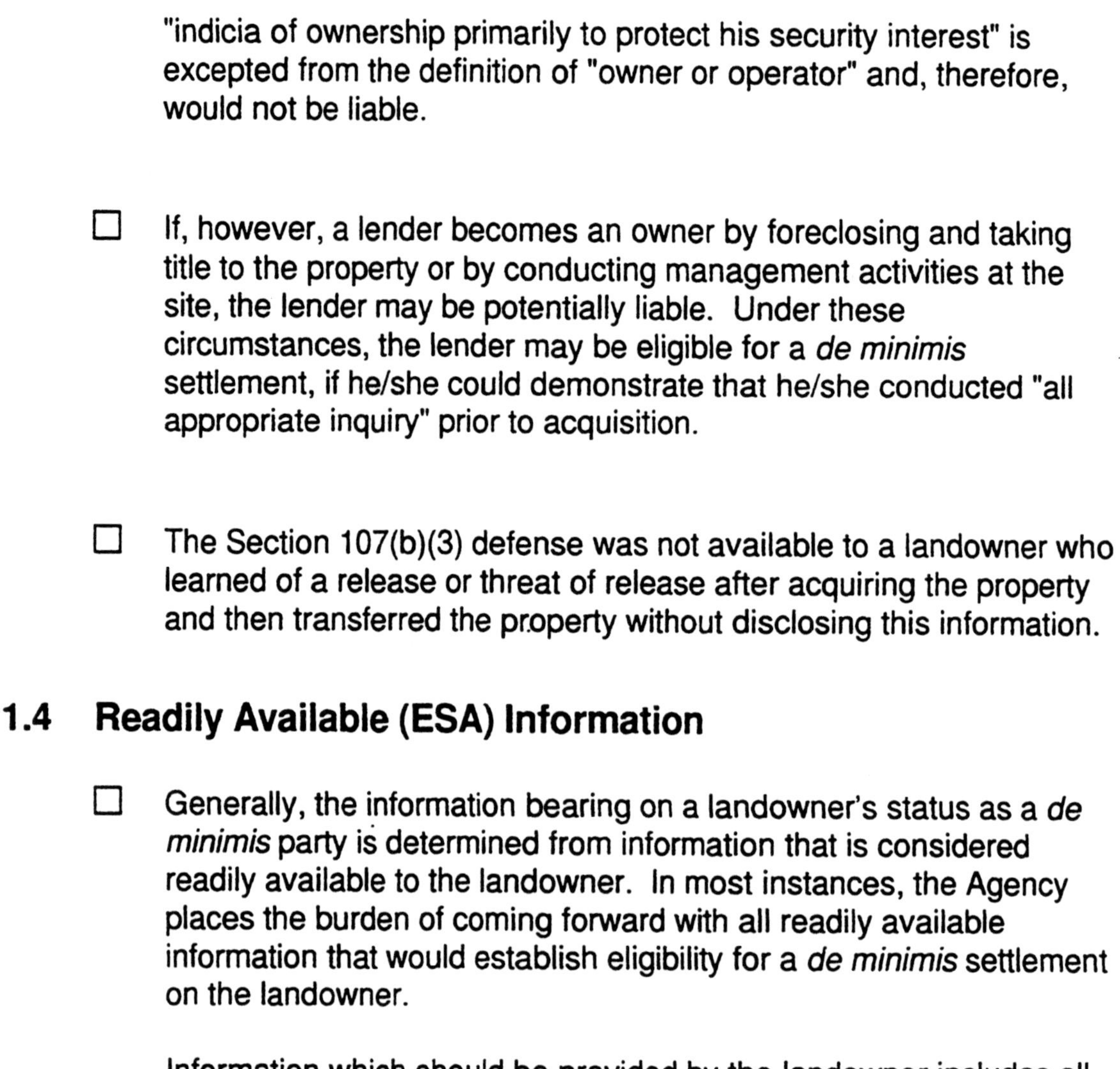

"indicia of ownership primarily to protect his security interest" is excepted from the definition of "owner or operator" and, therefore, would not be liable.

- ☐ If, however, a lender becomes an owner by foreclosing and taking title to the property or by conducting management activities at the site, the lender may be potentially liable. Under these circumstances, the lender may be eligible for a *de minimis* settlement, if he/she could demonstrate that he/she conducted "all appropriate inquiry" prior to acquisition.

- ☐ The Section 107(b)(3) defense was not available to a landowner who learned of a release or threat of release after acquiring the property and then transferred the property without disclosing this information.

1.4 Readily Available (ESA) Information

- ☐ Generally, the information bearing on a landowner's status as a *de minimis* party is determined from information that is considered readily available to the landowner. In most instances, the Agency places the burden of coming forward with all readily available information that would establish eligibility for a *de minimis* settlement on the landowner.

  Information which should be provided by the landowner includes all evidence relevant to the actual or constructive knowledge of the landowner at the time of acquisition, including all affirmative steps taken by the landowner to assess:

  - Site History (previous ownership and previous uses of the property)
  - Site Location and Description (information regarding the condition of the property at the time of purchase)
  - Environmental Hazards and Special Resources (all documentation and evidence of representations made at the time of sale regarding prior uses of the property and information regarding any specialized knowledge on the part of the landowner which may be relevant)
  - Findings, Summary, and Recommendations (relevant to the issues of whether the landowner exercised due care,

and whether the landowner contributed to the release or threat of release through any act or omission)

- ☐ This environmental site assessment information included:

  - Hazardous substances identified and the extent of knowledge regarding the substances
  - Abatement measures taken (if any)
  - Measures taken by the landowner to prevent acts of third parties which may have contributed to the release
  - Assurances that due care would be exercised

1.5 Innocent Landowner Status, Need for ESAs

- ☐ To be considered an *Innocent Landowner* under CERCLA, a purchaser has to prove that an environmental hazard was attributable solely to contamination which existed prior to the date of his/her acquisition. Herein lies the need for the Phase I ESA.

- ☐ The *due care requirement* embodied in Section107(b)(3) requires persons to exercise a degree of care, in obtaining, assessing, and providing that information which is reasonably available under the circumstances, in a written report (ESA). An excerpt from the text of CERCLA Section 107 that relates to ESAs is included in the Preface.

- ☐ The "information readily available to all" required by the agency consists of that information that a person knowledgeable of the sources can find. That "person" is the ESA inspector. Phase I ESAs conducted by third party independent professionals have become the easiest way to break into the environmental services field.

1.6 Types of ESAs

There are three types of "Phased Environmental Assessments." These are the :

- Phase I ESA;
- Phase II ESA; and
- Phase III ESA.

1.6.1 Phase I ESA

- ☐ Phase I is considered the research phase whereby data on current and past site operations is collected and analyzed to determine the "potential existence " of onsite environmental impairment.

1.6.2 Phase II ESA

- ☐ Phase II is considered the sampling phase whereby the "potential existence " of onsite environmental impairment has been confirmed, and the extents of such contamination must be decided. Many a firm has gone bankrupt with the assumption that they had the requisite skills and personnel to conduct a Phase II.

1.6.3 Phase III ESA

- ☐ Phase III is considered the remediation phase whereby the "types and extents" of onsite environmental impairment has been confirmed and abatement plans or a scope of work for site abatement has been prepared. As with the Phase II, this is no place for the novice inspector, and many a firm has gone bankrupt attempting to conduct their first Phase III.

The remainder of this book will focus on the Phase I ESA process. Generalized comments on the other types of ESAs are made in the text, and it bears repeating: Phase II and Phase III ESAs are much more complex than the Phase I ESA and should not be attempted by the novice inspector or firm. My company conducted over 100 Phase I ESAs prior to attempting our first Phase II.

1.7 Purpose of a Phase I ESA

- ☐ The primary purposes for conducting a Phase I ESA, then, are threefold:

  1) assess environmental hazards and special resource impacts emanating on and around properties on a case-by-case basis;

  2) ensure that all appropriate inquiry and due care requirements are met in the ESA process; and

3) give the client an accurate assessment, based upon readily available information, of any environmental impairment and potential liability issues that may pertain to the subject property.

1.7.1 Hazard Investigation Purpose

☐ The information on hazards which is presented in the Phase I Environmental Site Assessment (ESA) is used to evaluate a property's legal and financial liabilities for real estate transactions related to foreclosure, purchase, sale, loan workout, or seller refinancing.

1.7.2 Special Resource Investigation Purpose

☐ The purpose of identifying a special resource is to evaluate:

- the property's overall development potential,
- the associated market value, and
- the impact of applicable laws that restrict financial and other types of assistance for the future development of the site or require specific actions to be taken prior to the sale of the property.

1.8 Special Provisions

☐ There are several key requirements that must be understood prior to conducting the Phase I ESA. These issues are imperative to the successful completion of the ESA protocols outlined in this book.

1.8.1 Property Condition

☐ The ESA inspector should always restore the property to the condition in which it was found, with the exception of minor cuttings, sampling, and other materials removed for the purpose of investigation.

☐ The inspector should make every effort to minimize interference with property condition and operations while conducting investigations.

1.8.2 Attachments and Certifications

- ☐ In addition to the information provided in the appendix to the ESA report, the inspector should reiterate the Statement of Work (SOW) in the ESA report. The inspector and any subinspectors should affirmatively state that they are properly licensed and/or certified to do the work described in the SOW and can provide evidence of such licenses and certificates.

- ☐ All laboratories for asbestos and other sampling and analysis should be accredited and in compliance with applicable federal and state requirements.

- ☐ For new inspectors without the requisite experience, teaming is the solution. More on that later.

1.9 Confidentiality

- ☐ The inspector should ensure that his/her associates, employees, agents, and subinspectors do not disclose the report or any information contained therein to any person without the prior knowledge of the site representative, unless otherwise required by applicable law.

- ☐ The inspector should instruct each of his/her associates, employees, and independent inspectors as to the confidential nature of ESA reports and any information therein, and they should be required not to disclose such materials and information or otherwise act to breach the confidentiality of these materials or information.

1.10 ESA Updates

- ☐ If the inspector receives any information on the site after submitting the report, the inspector should immediately notify the property representative of the substance of such information and its potential impact on the subject site.

- ☐ Additional information should be mailed or delivered to the representative within two (2) to five (5) business days from the inspector's receipt of such information.

- ☐ If information related to an eminent and foreseeable hazard to public health and welfare is received post-submission, phone contact followed by same-day delivery, or overnight mail, is advisable.

- ☐ If the hazard is a known fire threat, it is best to notify local fire and health authorities along with the representative. If the hazard involves potential release of listed substances, EPA should be notified as well.

***

ESA Statement of Work (SOW)

The SOW is a project plan describing what services will be provided. Notwithstanding any previous Phase I ESA Statement of Work, the following requirements provide an outline for conducting Phase I ESAs. This outline is intended to provide consistency in your ESAs. Many agencies, financial institutions, environmental associations, and standard-setting entities (e.g. ASTM) have adopted their own SOWs for Phase I ESAs.

2.1 SOW Need

☐ You should always find out if your client requires a specific SOW prior to bidding. If none has been adopted, two options exist for the inspector that should meet or exceed his/her client's needs:

- develop an SOW by stating the specific tasks that will be conducted based upon the information presented in Sections 7 through 11 (see Sample Phase I ESA in Appendix A), or
- obtain a copy of the Phase I ESA SOW available from the American Society of Testing and Materials (ASTM).

☐ Note: the Phase I ESA SOW developed by ASTM may be obtained from:

ASTM
1916 Race Street
Philadelphia, Pennsylvania 19103
(215) 299-5585

☐ Whenever an SOW investigation requirement has not been met, the inspector should indicate this, and the reason the requirement was not met, in the summary and recommendations section of the report.

2.2 The Four Methods of Inquiry

- ☐ Typical methods of inquiry include, but are not limited to, the following:

  - Review of site background
  - Review of agency records
  - Interviews with persons knowledgeable about the site and surrounding and adjacent area
  - Site reconnaissance and investigation

- ☐ These steps are typically conducted simultaneously, and the results of each, used together, constitute the final ESA report.

2.3 Sampling

- ☐ Although the Phase I may allow for some sampling such as in the case of asbestos, invasive procedures or sampling are generally part of a Phase II SOW.

- ☐ If the inspector finds that certain investigations or sampling is needed, beyond any authorized at the time of contract award, the site owner, manager, or contracting representative should be notified in writing before beginning work. The inspector should never proceed with any type of site sampling until written approval from the responsible representative is received.

- ☐ If sampling is required, the inspector must usually be certified by a state regulatory agency. If the inspector does not have the needed certifications, it is usually permissible to subcontract the sampling and analysis tasks out to certified personnel or firms. A phone call to the state agency responsible for certifications may be all that is necessary to identify potential subcontractors.

***

Overview of ESA Format

The inspector's assessment report should address the objectives of an ESA as described in Section 1, "Purpose of a Phase I ESA." The following is a summary of the minimum contents of the report. The report should be supplemented as necessary by the inspector.

3.1 Scope of Work

- ☐ The report should include a brief discussion of the assumptions used in conducting the environmental assessment and in determining its scope.

- ☐ The inspector should include a discussion of any procedures not utilized under the ESA Statement of Work, with the rationale for omission.

- ☐ Note: For the purposes of this book the terms *Statement of Work, Scope of Work,* and *Scope of Services* are synonymous.

3.2 Property Information

- ☐ This section should clearly set forth the following information:

  - Property name and address (including city, county, state, and zip);
  - Name of owner, property manager, and/or site representative responsible for the contract;
  - Any unique property identifiers used by the contracting party (e.g. loan numbers, asset numbers); and

- The name and address of any private management firms or other third party contact offices responsible for the property.

3.3 Site Location and Description

☐ This section should contain information about the site, including:

- Legal description (including lot, block, subdivision) and street address;
- Type of property and current and prior use;
- Improvement(s) on property; and
- Owner(s), operator(s), manager(s), and occupant(s) on property.

3.3.1 Adjacent Sites

☐ The inspector should provide a summary of current and prior adjacent land uses, especially focusing on those which may pose hazards to the subject property.

☐ The inspector should briefly describe the number of properties which are managed by governmental, commercial petroleum entities (UST owners/operators), and other facilities which by the nature of their operations may pose a hazard to the subject site.

3.4 Site History

☐ This discussion should include a review of, and conclusions drawn from:

- A chain-of-title search
- Aerial photographs
- Local street directories
- Maps and other historical sources

☐ Significant information obtained from interviews with persons knowledgeable about the site should be included here.

3.5 Regulatory Review

- ☐ This section should include an evaluation of all records reviewed, including general public records, the environmental database search, and a summary of the results of all interviews and inquiries made of federal, state, or local regulatory authorities.

- ☐ A brief synopsis of the results of the environmental information database search as it impacts the property should also be provided.

3.6 Site Investigation and Review of Hazards

- ☐ This includes a site evaluation of:

  - Asbestos-Containing Materials (ACM)
  - Chemicals and Raw Materials
  - Geology
  - Hazardous Substances
  - Hydrology
  - Landfills
  - Lead-Based Paint
  - Lead in Drinking Water
  - Offsite ASTs and USTs
  - Onsite Aboveground Storage Tanks (ASTs)
  - Onsite Underground Storage Tanks (USTs)
  - Pits, Sumps, Drywells, and Catchbasins
  - Polychlorinated Biphenyls (PCBs)
  - Radon
  - Soils
  - Stormwater Drainage
  - Topography

3.7 Site Investigation and Review of Special Resources

☐ This includes a site evaluation of:

- Coastal Zones
- Covered Property
- Endangered Species
- Floodplains/Wild and Scenic Rivers
- Historic Property
- Natural Landmarks
- Recreational Areas
- Scientific Significance
- Sole-Source Aquifers
- Wetlands
- Wilderness Areas

3.8 Findings: Environmental Hazards and Special Resources

☐ This section should include, but not be limited to, a discussion of findings and conclusions, focusing on areas where there are potential or suspected environmental concerns.

☐ All findings and conclusions should be supported by, and referenced to, the preceding content of the ESA.

☐ All findings and conclusions should apply to the subject property, as well as to surrounding properties.

☐ When applicable, the inspector's findings should include a statement of the Permissible Exposure Level (PEL) and/or Threshold Tolerance for each hazard (as defined by federal, state, or local law), and a statement regarding the development potential of special resource areas.

☐ This section should include a discussion of any investigation requirement that was not satisfied, or information on any resource that was not available, and the reasons why the requirement was not satisfied or the information on any resource was not available.

☐ This section should include a discussion of each area of concern that is an actual or potential source of liability which needs further investigation.

☐ This section should include a discussion of actual or potential noncompliance issues with environmental laws, regulations or standards, including any potential impact on the future use of the site.

☐ This section should include recommendations on further investigations, if necessary, to evaluate whether contamination or special resource concerns exist at the site. These recommended investigations should be supported by the report's findings as well as the professional opinions of the Phase I inspector regarding the potential for environmental contamination, hazards or special resource value concerns at the site.

☐ This section should include recommendations for regulatory reporting which may be required by results of this investigation, including:

- Name of the agency, contact individual, telephone number, facsimile number (if available), street and mailing address
- Law or regulation which requires reporting
- Time periods during which such reports should be made
- Copy of reporting forms (if any)

3.9 Asbestos-Containing Materials (ACM) Recommendations

☐ The ranges of recommendations to be considered by the inspector for asbestos-containing materials should include response actions for removal, repair, encapsulation/enclosure, or operations and maintenance (O&M).

- ☐ The inspector should recommend the least costly response action that protects the health and safety of the building occupants for each homogeneous area.

- ☐ All recommendations must be consistent with the hazard assessment criteria in the Asbestos Hazard Emergency Response Act (AHERA).

- ☐ A chart showing the recommended response action for each homogeneous area should be provided.

- ☐ In developing any recommendation(s) for abatement procedures (repair, encapsulation, enclosure, or removal), the inspector should recommend limited replacement and patching for repair of damaged ACM and the specific type of encapsulation method for each ACM.

- ☐ If removal of friable, damaged asbestos is the recommended response action, the inspector should:

    - Identify permits required
    - Identify proper techniques for air monitoring and disposal
    - Identify equipment to be used
    - Estimate building material replacement costs
    - Estimate other relevant engineering costs
    - Estimate length of time required to complete the recommended response action(s).

3.9.1 ACM Abatement Costs

- ☐ The inspector should present cost estimates for the recommended response action(s), along with recommendations on other actions necessary to comply with any federal, state, or local law, ordinance, regulation, or permit requirements/restrictions which may be applicable to the property in connection with identified environmental hazards or special resources.

3.10 Warrants

☐ As defined by *Black's Law Dictionary*, to warrant is to promise that a certain fact or statement of facts, in relation to the subject matter, is, or shall be, as it is represented to be. In ESAs, the Warrant section is used as a promise that the facts presented in the study were true as of the date of the investigation.

☐ This section is extremely important to the inspector, because without it certain facts of the ESA may be implied without limitation (e.g. the condition of subsurface features). This presumption would place the inspector at risk for events that occur at a site after his/her site inspection tasks were completed. The format of this section is fairly typical and should, at a minimum, include:

- An exclusionary clause making the warrant the only one in effect
- A limitation of liability for conclusions based on data obtained from government agencies and third parties
- A statement that reasonable care and professionalism were utilized (quality assurance)
- A statement that the ESA relates to findings compiled during a specific time and place
- An acknowledgment that changes are inevitable
- An exclusive-use clause, limiting the ESA for the inspector's and client's use only
- A limitation on responsibility for independent conclusions and interpretations of the ESA

3.11 ESA Inspectors

☐ The name and title of all persons who conducted the ESA, and the indication of which of these persons was the project manager for the study and report, should be included in the ESA. This data is typically placed in one or more of the following ESA sections: cover letter, executive summary, title page or appendices.

***

Section 4

Environmental Databases

Environmental database reports search federal, state, and local databases and lists to identify potential hazards to the site emanating from onsite and offsite sources within prescribed search distances.

4.1 Regulatory Database Searches

- ☐ The environmental database search should cover a number of regulatory databases and lists.

- ☐ The environmental database search should cover governmental agency records pertaining not just to the property, but also pertaining to properties within the Minimum Search Distance (see definition in Glossary) in order to help assess the likelihood of problems from migrating environmental hazardous substances.

- ☐ The Phase I inspector should analyze and interpret the data obtained through the database search to conduct his/her analysis of the site, and to develop site reconnaissance priorities and recommendations in the ESA.

- ☐ The environmental database report may be prepared by an environmental database vendor or the Phase I inspector.

- ☐ The environmental database report should always be accompanied by a quick-reference summary sheet of database findings (see Appendix A).

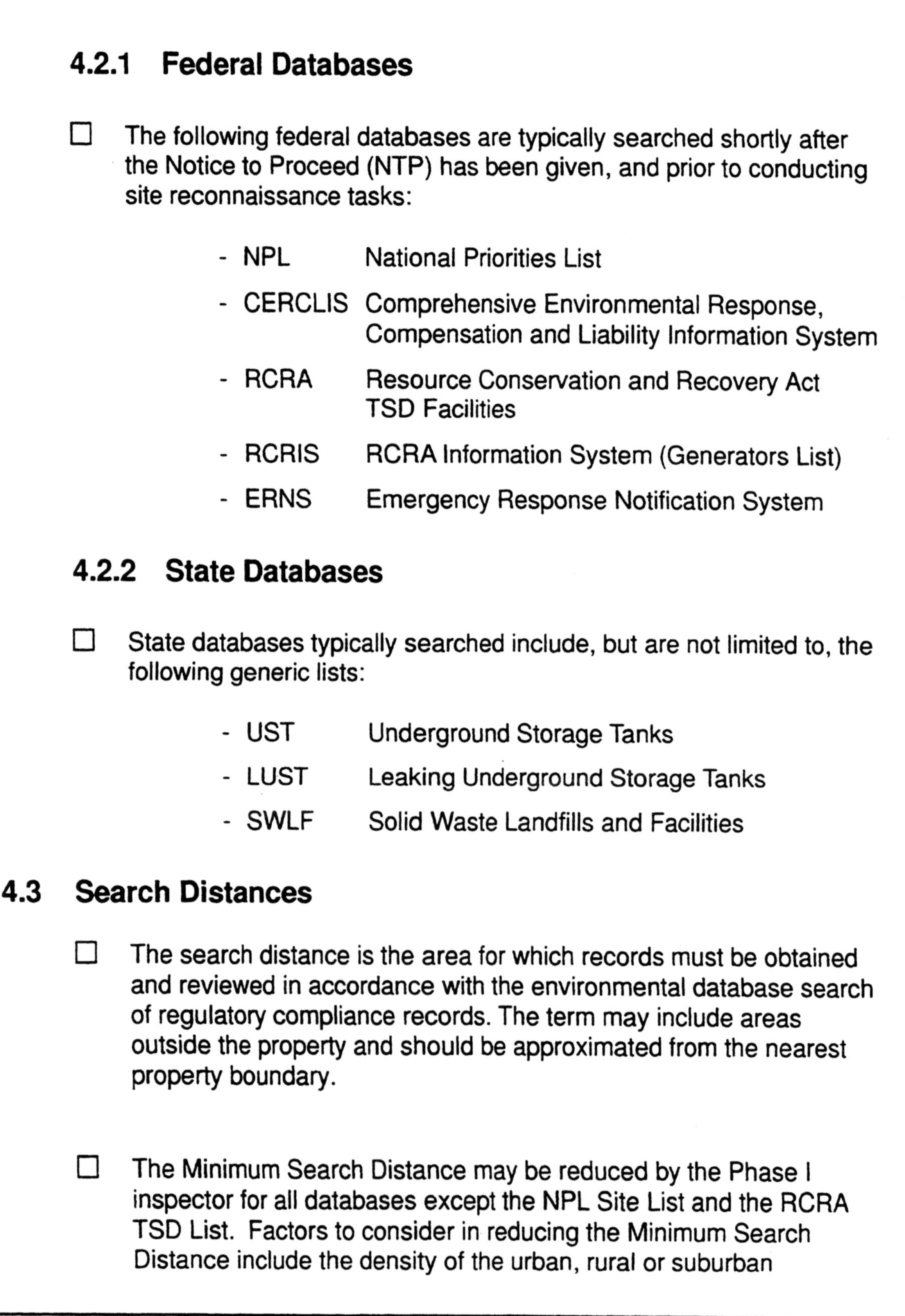

4.2 Typical Databases Searched

4.2.1 Federal Databases

☐ The following federal databases are typically searched shortly after the Notice to Proceed (NTP) has been given, and prior to conducting site reconnaissance tasks:

- NPL — National Priorities List
- CERCLIS — Comprehensive Environmental Response, Compensation and Liability Information System
- RCRA — Resource Conservation and Recovery Act TSD Facilities
- RCRIS — RCRA Information System (Generators List)
- ERNS — Emergency Response Notification System

4.2.2 State Databases

☐ State databases typically searched include, but are not limited to, the following generic lists:

- UST — Underground Storage Tanks
- LUST — Leaking Underground Storage Tanks
- SWLF — Solid Waste Landfills and Facilities

4.3 Search Distances

☐ The search distance is the area for which records must be obtained and reviewed in accordance with the environmental database search of regulatory compliance records. The term may include areas outside the property and should be approximated from the nearest property boundary.

☐ The Minimum Search Distance may be reduced by the Phase I inspector for all databases except the NPL Site List and the RCRA TSD List. Factors to consider in reducing the Minimum Search Distance include the density of the urban, rural or suburban surroundings in which the property is located; and the distance that a hazardous substance is likely to migrate based upon local geologic or hydrogeologic conditions. In considering search distances for USTs, an upgradient site is unlikely to be impacted by any downgradient USTs. In this instance an inspector can restrict his/her search on the upgradient side of a site.

- ☐ Changing search distances and/or directions typically require that the inspector have in-depth information on local site history and subsurface conditions.

4.4 Database Vendors

- ☐ If the Phase I inspector retains a firm to conduct an environmental database search, the search distance for federal and state record sources should be in accordance with the "Minimum Search Distance" (see Section 4.3 above).

***

ESA Report Appendices

All ESAs require certain supporting documents and documentation to validate the report's findings and recommendations. The environmental data to be included in a report's appendices are a very important part of the ESA. Often times comments from a client's reviewer (attorney or employee with environmental expertise) center on a lack of a certain appendix that will validate a finding or conclusion in the study.

5.1 ESA Supporting Documents

☐ The appendices should include, but should not be limited to, the following information, as applicable:

- The ESA cover letter
- All regulatory databases searched
- Abstract or synopsis of title or other written title search materials
- Aerial photographs
- Maps and other data, including a site plan drawing prepared by the inspector
- Site photographs with labels (originals or copies of 35-millimeter color photographs of the site) including orientation to road frontage and/or compass direction or geographic features
- Copies of any permit applications or permits issued to the property
- A list of agencies contacted for records search and names of persons contacted
- Records of agency interviews/conversations via telephone or personal visits

- List of persons interviewed at or in the vicinity of the site, and records of conversations with these persons, including the date of the conversation and identity of individuals present during the conversation. The relationship of each person to the property should also be specified
- Reports or completed notification/registration forms submitted to state or local regulatory authorities
- Violation notices from federal, state or local regulatory authorities
- Copies of written inquiries and responses to and from federal, state, or local agencies concerning the existence of hazardous substances, hazardous wastes, or special resources
- Asbestos and lead-based paint surveys
- Copies of all required licenses and certifications for all persons who conducted the ESA
- Reference documents, such as soil or hydrology studies
- Copies of completed environmental checklists
- Other ESA data references

5.2 The ESA Cover Letter

☐ Every ESA submission should be accompanied by a cover letter based on what I call the SF2RC format (S for synopsis, F for findings, R for recommendations, R for risk, and C for contact).

5.2.1 Project Synopsis

☐ The project synopsis is an overview of all the vital project data required for someone uninformed about the site to get up to speed immediately on the following aspects:

- Property type;
- Inspection date and inspector;
- Any site identification and contract numbers; and
- What Statement of Work (SOW) was used, etc.

5.2.2 Findings

☐ The findings included in the ESA report should be copied verbatim and placed in the cover letter. Each separate finding should be made its own paragraph, and numbered or otherwise highlighted to stand out.

5.2.3 Recommendations Reiterated

☐ Recommendations should be copied verbatim from the ESA report and written in a manner that corresponds to each finding. In the event a finding was made but no adverse impact is anticipated, the recommendation should be "No impact anticipated, because... [full rationale for the no action/no impact conclusion must be given]..., and no additional work required at this time."

5.2.4 Risk Rating

☐ For findings which may pose adverse impacts to the site, the inspector should assess one of the following risk ratings: Little or No Risk, Low Risk, Low to Moderate Risk, Moderate Risk, Moderate to High Risk, or High Risk. These ratings are derived from the opinion of the inspector and should be based upon the severity of the potential environmental threat(s). The lowest rating should be given to offsite hazards that pose no threats to the site, and the highest ranking should be given to onsite and offsite sources that pose eminent threats to public health, welfare, and property.

5.2.5 Contact Data

☐ Readers of the ESA and cover letter should always be able to contact the firm and/or individual who conducted the study for additional information, so the phone and FAX numbers, address, inspector's name, and firm name should always be on the ESA cover letter.

5.3 Maps and Map Presentations

☐ Some of the most important exhibits found in an ESA are maps. Maps of various kinds should always be collected, analyzed, and included in the study as appropriate.

5.3.1 Map Templates

☐ Present maps in an ESA standardized template. A map frame should be used that immediately orients the reader with a north arrow, a map scale, a map date, a map title, and the name of the person or agency that prepared the map.

5.3.2 Identifying Site Locations

☐ There are various ways to mark sites on maps. Stock graphics of arrows can be purchased as stick-ons from many art and office supply stores. I recommend using an inexpensive software package to print out a full page of north arrows, site location arrows, etc., cutting them out as you need them, and using gluestick to adhere them to your maps. The same can be done with street names, unique site features, identifying adjacent properties, etc.

5.3.3 Site Map Resources

☐ Preparing site maps is easier than one might think, because the key is to find an existing map of the subject area and adapt it to the standard template. When we first began conducting ESAs several years ago, we were tempted to draw every site map in our field notes. But after conducting 70 or so ESAs it became evident that most municipal and all county governments maintained various types of maps that are either given away free for the asking or reproduced for a modest duplication fee.

5.3.4 USGS Quads

☐ The greatest source of maps is the United States Geologic Survey. The USGS has 7.5 Minute Topographic (Quadrangle) maps starting at about $2.50, as well as a myriad of types of maps. The only concern is that the scale of these products is usually very large, so that while these maps fulfill the needs for topographic review and site vicinity issues, they cannot be used as site specific maps.

5.4 Sampling and Analysis

☐ Appendices for radon, asbestos, and/or lead-based paint, if applicable, should include:

- Technical memoranda on field activities
- Analytical data and quality assurance/quality control (QA/QC) evaluation results
- Photographs (35-millimeter color) and a log of all samples and suspect asbestos-containing building materials

☐ Floor plans and building blueprints should be included when available. These plans should show the locations of all suspect materials that were reported to contain asbestos, and provide a graphic presentation of sample locations and sampling grids.

☐ Sampling plans, with a table describing all suspect and known ACM, should be included that provide:

- A description of the material
- Location of the material
- The types and amounts of ACMs
- The conditions of the material, as determined in accordance with the guidelines established under AHERA (friable, nonfriable, etc.)

☐ A list of any inaccessible areas, and a summary of the impact of these inaccessible areas on the findings of the inspection, should be included. Other data to consider when conducting specialized surveys includes:

- Health and safety plan for asbestos inspectors
- Chain of custody forms
- Current certifications of asbestos inspector, asbestos management planner
- Current National Voluntary Laboratory Accreditation Program (NVLAP) certificates from the analytical laboratories

5.5 Other ESA Data References

☐ Just about all federal, state and local governments prepare various specialized studies and reports that contain valuable information for completing a Phase I ESA. (See Section 5.5.1)

- ☐ Besides governmental entities, there are a myriad of associations in the environmental and related fields, such as engineering, architecture and planning, that also conduct studies and publish reports.

5.5.1 Plans, Studies, Reports, and Permits

- ☐ When the subject site is in a medium-to-large urbanized area, the typical types of studies that may be found include:

  - Air quality permits
  - Building permits
  - City general plans or comprehensive plans
  - Conservation plans
  - Earthquake hazards surveys and professional papers
  - Environmental assessments and groundwater monitoring reports for nearby sites
  - Infrastructure development plans
  - Open space plans
  - Remedial action/final closure reports for UST sites
  - Soils investigations
  - Special district plans
  - Tax assessor's data and maps
  - Other related information

5.6 Local Governments

- ☐ Many of these sources of data are prepared by local governments focusing on specific areas. In that case, a single study may contain numerous maps and details that may be used as exhibits to validate the inspector's work.

***

The ESA Process

A systematic approach is always best when conducting Phase I ESAs. The following nine-task approach has been used extensively by the author to complete over 100 Phase I ESAs. Use it as a guide, adapting specific aspects as required by a particular Statement of Work. Utilizing the four methods of inquiry for ESAs and this nine-step approach that includes administrative responsibilities pre- and post-inspection, an inspector should be able to complete a Phase I ESA within two to three weeks.

6.1 The ESA Process

☐ Once a contract is awarded, conducting ESAs is a very methodical process requiring a minimum of nine distinct tasks. These tasks are:

- Preparing bids and letter agreements
- The analysis of existing site data and conditions
- Site history analysis
- Regulatory agency and database analyses
- Site interviews
- Site reconnaissance
- ESA report preparation
- ESA report submission
- ESA review

☐ Preparing bids and letter agreements, ESA report preparation and submission, and ESA review are actually preliminary and secondary steps and are discussed later in this section.

- ☐ The analysis of existing site data and conditions, site history analysis, regulatory agency and database analyses, site interviews, and site reconnaissance tasks constitute the actual ESA process and will be discussed in Sections 7 through 11.

6.2 Bids, Letters and Agreements

- ☐ The inspector should submit a standardized letter agreement showing the proposed cost for conducting the ESA. This letter should contain information on the fee for conducting a standard Phase I ESA, as well as for additional procedures and/or tests which may be required of the inspector for specific types of sites (e.g. those requiring asbestos or lead-based paint surveys).

- ☐ The bid letter should specifically identify the inspector(s) proposed for use and their relevant qualifications in accordance with the requirements of the ESA.

- ☐ One letter can accomplish the task of introducing your skills and abilities, providing a fee quote for a specific property, and providing a signatory section that constitutes acceptance of the bid and notice to proceed.

6.3 ESA Report Preparation

- ☐ Computer literacy is a must for conducting Phase I ESAs. Client names and addresses can easily be merged into boilerplate cover sheets, labels, and related correspondence, and report modifications can be made easily.

- ☐ It is not necessary to have a copy machine, as it is just as efficient to use a quick copy-type service that is located near, preferably next door to, a post office.

- ☐ Once the originals are produced, then copying, binding, and mailing can all occur within a minimal amount of time.

- ☐ Typically, two or three copies of the report should be submitted to the owner, manager, or other site representative, and one complete copy should always be retained by the inspector. The originals used for duplication should be filed in a safe place.

6.4 ESA Report Submission

- ☐ The report should be submitted no later than the agreed upon number of business days after the contract award. Based on the author's past experience, a straightforward ESA on relatively hazard-free land can be completed in 8 working days (in a rush situation). However, 20- to 30-day time frames are considered typical.

- ☐ Each copy of the ESA should include a copy of the cover page and title page signed by the inspector (see sample in Appendix A). The author suggests each copy have an original signature in blue ink, but experience shows that some clients only require one original report.

- ☐ The invoice should also be submitted with the ESA report. The invoice should contain the following:

    - Inspector's or firm's name, address, and phone number
    - Taxpayer Identification Number
    - Project name
    - Contract number
    - Brief description of services rendered
    - Date of project completion
    - Date of invoice
    - Breakdown of fees

6.5 ESA Report Review

- ☐ The site representative may ask for a third-party legal review, from an outside attorney, following his/her receipt of the study. Under these circumstances, that representative should simultaneously transmit any resulting comments (technical and/or legal) to the inspector, for immediate clarification.

- ☐ If comments are lengthy, an addendum addressing all concerns should be prepared by the inspector and submitted. It is then the representative's responsibility to attach a copy of the addendum to all copies of the ESA. In most cases this is sufficient.

- ☐ If additional revisions of the report are required, or if an addendum becomes very lengthy, the inspector should incorporate all changes into a revised ESA report (after the representative has approved the addendums, or has had them approved) and resubmit the entire ESA study.

- ☐ Rewriting an entire ESA into a Final may be required on your first few ESAs until the process is mastered, and you should plan your time accordingly. Initially, you might consider writing a Draft ESA, and following it up with a Final if needed. This issue must be considered in all your ESA bids. The author always assumed there would be a Draft, then a Final. By not indicating Draft on the cover page, or in the study, many studies are accepted as Finals because they meet the client's environmental data needs. This is good for the inspector because it saves him/her time, reproduction, and mailing costs.

***

Step One: Current Site Data

Once awarded the contract and given notice to proceed, the ESA inspector should review all available prior environmental studies, and site documentation, including: previous ESAs, real estate appraisals, and related environmental checklists prepared for the property. This information should always be considered when preparing the ESA.

7.1 Analysis of Existing Site Conditions

☐ Existing ESAs and related data should be supplied to the inspector by the property owner, manager, or other site representative. If this data is not supplied, the inspector must ask for it.

☐ The inspector should always start the ESA process by identifying property by legal description (including lot, block and subdivision), street address, municipality, county, state, and ZIP Code. This information should be provided by the client's representative.

☐ The next issue is to identify the basic site data, including but not limited to:

- Property type (e.g. commercial office building, multifamily, industrial facility, undeveloped land, etc.)
- Property size (acres)
- Number of building structures on property
- Building size (in square feet) and the number of stories in the building
- Building age
- Current occupancy status

- ☐ Much of the existing or current site data, listed above, should be provided by the client or his/her representative.

7.1.1 Identifying Uses of the Property

- ☐ The inspector should indicate the names of:

  - Owner(s)
  - Operator(s)
  - Manager(s)
  - Occupant(s)

- ☐ By defining the above, the inspector gains insight into the past nature of onsite businesses and their operations.

7.2 Site Vicinity

- ☐ For hazards, the inspector should identify present land uses on adjacent properties that may pose an adverse impact on the subject site. Red flags should raise in the mind of the inspector if, for example, any facilities in close proximity involve the following:

  - Manufacture
  - Generation
  - Use
  - Storage and/or disposal of hazardous substances

7.2.1 Site Vicinity Impacts

- ☐ In determining potential hazard impacts from facilities described above, the factors to consider include:

  - The extent to which information is readily available, useful, accurate and updated
  - Hydrogeologic/geologic conditions of the property that may indicate a high probability of hazardous substance migration to the property
  - How recently local development has taken place

- Information obtained from interviews
- The extent to which history and/or uses of properties in the site vicinity are generally researched

7.3 Sites Greater Than 50 Acres

☐ For partially or undeveloped sites greater than 50 acres in size, the inspector should identify any properties adjacent to, or contiguous with, the site which are managed by a governmental agency for:

- Wildlife preservation
- Open space conservation
- Recreational uses
- Historical conservation
- Cultural conservation
- Natural resource conservation

☐ The 50-acre size threshold is imposed by various statutes and federal agencies.

7.4 Existing Documentation: Environmental Database Reports

☐ The property owner or manager is responsible for supplying the inspector with any prior environmental database searches conducted for the site. If no searches have been conducted, the Phase I inspector is to proceed with his/her own database search at the time of receiving a notice to proceed (NTP) with the Phase I ESA.

☐ Please note: If an environmental database search was previously done and the resulting report is less than ninety (90) days old, the inspector should be able to utilize this report instead of ordering a new one. Client preference, report quality and validity should be the determining factors as to whether an additional search is required.

7.5 Existing Documentation: Special Resources

☐ If a prior ESA has been conducted, the property owner/manager is responsible for supplying the inspector with the results of the Special

Resources review simultaneously with notice to proceed with the Phase I ESA.

☐ If a special resources review was conducted, the inspector should utilize this information.

☐ If a special resources review was not conducted, and the Phase I ESA inspector suspects that due to the nature, size, and/or location of the site, these resources may be present on or near the property, the Phase I ESA inspector is advised to proceed with the implementation of that process.

☐ Special resources data is typically maintained by the:

- Advisory Council on Historic Preservation (ACHP)
- State Historic Preservation Officer (SHPO)
- The Nature Conservancy (TNC)
- Local/state historic preservation entities

***

Step Two: Site History

The inspector should identify historic uses of the property. A reasonable attempt should be made to identify readily available information on site usage over the past 50 years, or back until a time when the property and area were undeveloped.

8.1 Title Search/Title Review

☐ Conduct a minimum 50-year chain-of-title search of the site and obtain a title abstract. This can be done by researching deed books at the county clerk and recorders office, for the county in which the property is located.

☐ As an alternative, have a title company provide a certificate of title abstract which shows the chain-of-title.

☐ Report on any documents which may indicate a basis for potential contamination of the site by hazardous substances.

☐ Report on any conditions, restrictions, and/or easements affecting any special resource area or the management of such area on the site.

8.2 Aerial Photographs

☐ Secure, review, interpret, and provide summaries of aerial photographs of the site and surrounding and adjacent areas. Many local and county planning or public works departments maintain a file of aerial photographs, and private vendors are also available.

- ☐ The inspector should obtain all reasonably available photographs. The optimum would be to obtain photographs dating back 50 years, and in intervals of 10-15 years. Often times the range and number of aerial photos available are less than optimal.

- ☐ At a minimum, the frequency of aerials should represent periods with evidence of site activity or site development.

8.3 Maps and Other Data

- ☐ Secure, review, interpret and provide summaries of documents and maps regarding geologic, hydrologic and coastal zone conditions at the site and at the property surrounding the site. These maps include:

    - Archival maps
    - City maps
    - Sanborne maps
    - Floodplain maps
    - Land use maps (such as USGS and Sanborne Fire Insurance maps)
    - Site maps
    - Topographic maps
    - Utility maps
    - Wetland maps
    - Zoning maps
    - Others, including building blueprints or construction documents

8.4 Special Resources

- ☐ Review and interpret data maintained by the following agencies: National Park Service (NPS), the Advisory Council on Historic Preservation (ACHP), the State Historic Preservation Office (SHPO), the U.S. Fish and Wildlife Service (USFWS), and others. The pertinent data includes but is not limited to the:

- Endangered species list, including those species proposed for listing
- Environmental Protection Agency (EPA) list of designated sole-source aquifers, including those proposed for designation
- List of archaeological and historic properties listed or eligible for listing within the appropriate state
- List of wild, scenic and recreational rivers
- List of national trails and areas included in the appropriate state comprehensive outdoor recreation plan
- List of wilderness areas
- Map of the National Wilderness Preservation System
- National Registry of Natural Landmarks

***

Step Three: Governmental Records

Review and interpret federal, state, and local public records to assess the site, as well as areas surrounding the site. In medium-to-large urbanized areas, copies of most of these records can be obtained by visiting municipal and county government offices.

9.1 General Public Records

☐ The inspector's review should include but not be limited to the following:

- City engineering plans that indicate easements for water, sewer and electric utilities
- Property tax records and appraisal or assessment records
- Zoning/land use records
- Building permits and demolition permits
- Local and state trial court records for liens affecting the site
- Historical and archaeological records

9.2 Regulatory Compliance Records

☐ Review the vendor supplied, or inspector prepared, environmental database search report. This database search should cover a number of federal and state regulatory databases and lists (see Section 4, Environmental Databases).

☐ The environmental database search should cover governmental agency records pertaining not just to the property, but also to properties within a Minimum Search Distance (see Glossary), in order to help assess the likelihood of problems from migrating hazardous substances.

9.3 Agency Inquiries

- ☐ Conduct inquiries by telephone, in writing, or through visits with appropriate state, county, and municipal offices for information on existing and suspect environmental hazards and on special resources that was not available from the environmental database search.

- ☐ Personal visits to agencies are always more preferable, because you will inevitably collect other ESA data while visiting an agency, or you will learn that a sister agency with more data than you need is right across the hall.

- ☐ All verbal inquiries should be documented in a written record of conversations.

- ☐ Both records of verbal inquiries and written requests and responses should be included in the ESA report as part of the appendix.

9.3.1 For Environmental Hazards

- ☐ Obtain and review information regarding environmental violations, incidents, and/or status of enforcement actions/consent decrees, at the site and at property within a one-mile radius that may pose an environmental threat to the site. Contacts include, but are not limited to, the:

  - Emergency response agencies
  - EPA Regional Office
  - State and local environmental agencies
  - State and local health authorities
  - Local fire department
  - Local water and sewer authorities
  - Occupational Safety and Health Administration (OSHA) office
  - Sewer, water, and sanitation districts

- Water quality control departments

9.3.2 For Special Resources

☐ Contact applicable federal, state, and local regulatory agencies for information regarding regulatory requirements that should be met if special resources are present. Contacts include, but are not limited to, the:

- Local municipality or FEMA (if the property is situated in a 100-year floodplain)
- National Park Service or other agency responsible for administering a Wild and Scenic River or Wilderness Area
- Office of Groundwater for the Environmental Protection Agency (if the site is within the boundaries of a sole-source aquifer)
- Soil Conservation Service county office (for wetlands)
- State Coastal Zone Management Agency (if the site is within a coastal zone)
- State Historic Preservation Officer (if the site has any archaeological or historic significance)
- U.S. Army Corps of Engineers District Office (for wetlands)
- U.S. Fish and Wildlife Service (if endangered species or their habitat is present, or if the property is within a unit of the Coastal Barrier Resources System)

☐ Indicate in the summary and recommendations section of the report what federal, state, and local requirements may be special resources identified onsite.

9.4 Compliance Information Objectives

☐ Data to be obtained through regulatory compliance inquiries includes but is not limited to:

- Determining the status of federal, state, or local environmental permits or violations.

- Indicating in the report whether additional permitting is required under applicable regulations or laws for environmental hazards and special resources.
- Identifying corrective actions, restoration, or remediation planned, currently taking place or completed.
- Obtaining documents describing corrective action, restoration, or remediation plans or actions.
- Discovering and reporting hazardous substances.
- Obtaining and reviewing copies of Material Safety Data Sheets (MSDS).

***

Step Four: Interviews

Always interview individuals familiar with the site to determine the historic land use activities at the site.

10.1 Immediate Site

- ☐ Interview individuals familiar with the site to determine the historic land use activities at the site. Such individuals may include but are not limited to:

  - Present and former property owners
  - Employees
  - Key onsite managers
  - Occupants

- ☐ Identify possible hazardous substances used or released, waste streams, and prior use and ownership of the site and facilities.

- ☐ Maintain names, addresses, and telephone numbers of all persons interviewed.

10.2 Surrounding Area

- ☐ Interview persons at the immediate site to determine the historic land use activities in the area surrounding the site.

- ☐ Interview adjoining and/or hydrologically upgradient property owners (or other persons, such as occupants or key property managers having knowledge of the property in question) within a surrounding area that may affect the site (up to one mile, if identified in an environmental database) to obtain information about historic land use activities and conditions.

- ☐ The purpose of these interviews is to identify whether any environmental issues exist, including but not limited to:

    - Leaking tanks
    - Chemical spills
    - Authorized and unauthorized disposal activities
    - Hazardous releases

- ☐ The inspector's objective is to determine whether such incidents have affected the subject property in the past, or are currently negatively impacting the site.

- ☐ Interviews should be conducted with persons knowledgeable of the site vicinity, such as:

    - Real estate salespersons
    - Appraisers
    - County assessor
    - County land office personnel
    - Zoning commissioners, etc.
    - Others, whom in the normal course of their work may be aware of such issues

- ☐ The inspector's objective is to identify operations currently conducted on properties in the vicinity of the site.

- ☐ Maintaining records of names, addresses, and telephone numbers of all persons interviewed is extremely important.

***

Step Five: Site Reconnaissance

While the ESA inspector performs a visual inspection and walk-through of the site, he or she is observing and recording the distinct boundaries of the site, and improvements on it, with photographs, map directionals, and narrative notes for later presentations.

11.1 Visual Inspections of the Subject Site

Items of concern include, but are not limited to, the identification of:

- Sources of potable water for the property
- Oil, gas or disposal tanks or drums on the site
- Natural waterways
- The presence and condition of surface water discharges, paying special attention to discolored flowing or ponded waters
- Areas of stressed vegetation
- Indications of liquid or solid waste dumping or disposal
- Indications of ASTs and USTs
- Evidence of groundwater wells, cisterns, or septic tanks
- Abnormal odors associated with the site
- The presence of unnatural fill material or soil grading
- Whether or not special resources exist on the property
- Other indications of environmental and/or hazardous situations

11.2 Visual Inspections of the Site Vicinity

☐ The inspector must also conduct visual inspections of the site vicinity.

- ☐ Report in detail whether there are any activities that may use, generate, or store hazardous substances or hazardous wastes at the site and on properties surrounding the site.

- ☐ Note whether these activities are on properties upgradient or downgradient to the subject properties and discuss the potential for groundwater contamination.

11.3 Undeveloped Sites

- ☐ For sites having undeveloped land greater than 50 acres, the inspector should evaluate properties adjacent to or contiguous with the site, and report whether they are managed or regulated by a governmental agency (federal, state, local) for:

  - Wildlife refuges
  - Sanctuaries
  - Open space
  - Recreational activities
  - Historic preservation
  - Cultural resource preservation
  - Other natural resource conservation purposes

11.4 Site Photography

- ☐ All photographs should be original or photocopied 35-millimeter color prints of business quality or better.

- ☐ The inspector should adopt a methodical photographic approach and utilize the same system for photographing each ESA assignment (see below).

- ☐ Typically, the inspector will need to take a minimum of about 13 to 18 photos total for each site.

11.4.1 Exterior Site Photos

☐ A simple approach may be to always shoot the perimeter of a site in the following sequence:

- North facade or north side of site from the "north of site facing south" vantage point
- Immediately turn around and take the "adjacent north property" photo from the same vantage point
- Then move to the east facade or east side of site and take a photo from the "east of site facing west" vantage point
- Immediately turn around and shoot the "adjacent site east" from the same vantage point
- Move to the south and west sides of the site and repeat the same process as above

11.4.2 Exterior Hazard and Special Resource Photos

☐ Next look for any environmental hazards or special resources on or immediately adjacent to the subject site and take photos of them, noting their location on a field diagram or map. Typical features to look for include:

- Stained grasses
- Discarded drums
- Transformers
- Historic, cultural, or recreational resources
- USTs and ASTs
- Industrial operations and facilities
- Government facilities that store hazardous materials, such as motor pools
- Debris and illegal dumping activities, etc.

11.4.3 Interior Photos

☐ If buildings and structures are onsite, it is recommended that just prior to entering any building, a relatively close-up photo of the main entry with a visible address, name, or other identifying sign be taken first.

- ☐ If photos are taken sequentially following the above practice, it will help you organize photos by structure later.

- ☐ Upon entering the structure, take a photo of the first main feature that can be seen from the entryway, then take a few steps and immediately turn around to shoot the interior of the entryway. This practice should be conducted in each main area of the facility.

- ☐ Photos of dilapidated materials should be taken on a "worst case" basis, focusing on any structural condition that may present a hazard to public health and welfare first, then on materials which may be considered asbestos-containing or paints which are peeling and may contain lead.

***

Environmental Hazards Overview

Environmental hazards have the potential to pose health or safety risks to workers, residents, and the general public. These hazards may decrease the value of a subject site due to the high costs of cleanup associated with bringing the site into compliance with governmental regulations.

12.1 Environmental Hazards

☐ Based upon maps, aerial photographs, a title search, adjacent land ownership, and site reconnaissance, the inspector should ascertain whether the subject site has been, or is being, impacted by environmental hazards, such as:

- Underground storage tanks (USTs)
- Aboveground storage tanks (ASTs)
- Geologic hazards
- Soil hazards
- Topographic hazards
- Hydrologic hazards
- Hazardous substances (i.e. pesticides and other chemical contaminants)
- Asbestos-containing materials (ACMs)
- Lead-based paint/Lead in drinking water
- Polychlorinated biphenyls (PCBs)
- Landfills, pits, sumps, disposal sites, etc.
- Radon

12.2 Environmental Hazard Information Requests

- ☐ When requesting environmental hazard verifications from government agencies, the inspector should order and review an environmental database first.

- ☐ If issues arise, the inspector should then schedule the site reconnaissance to verify the database findings.

- ☐ If the hazardous issue is related to USTs or PCBs, the inspector should make phone contact with the appropriate state or public service company.

- ☐ If the hazardous issue is related to chemical spills, the local fire department is a good place to start.

- ☐ If the hazardous issue is related to any situation which would require Phase II sampling and analysis, the inspector should immediately contact the client for authorization to proceed. Examples of Phase II sampling situations include:

    - Soil testing
    - Well installation
    - Water testing
    - Asbestos surveys
    - Lead-based paint surveys
    - Radon surveys

12.2.1 Agency Environmental Hazard Information Requests

- ☐ Many regulatory agencies at the state level require a formal written request, accompanied by a service fee, prior to conducting a hazard review.

- ☐ Others require that the inspector schedule an appointment to come into their offices and review the files personally.

☐ If you will be conducting many ESAs, it is in your best interest to establish a working relationship with agency personnel early on. This will help you learn how their files and databases are arranged and accessed and get you over a major hurdle in the ESA process.

☐ When formal written requests are required, the inspector should include:

- An 8½ x 11 copy of a street level map of the area, with the subject site's boundaries outlined
- The property's name and address, including city, county and state
- The legal description including lot, block, township, range and section, as applicable (this information can be obtained from the client, property appraisals, or the county or city assessor)
- A general description of the property's physical characteristics, including acreage (this can be obtained from a site visit)

***

Underground and Aboveground Storage Tanks

The inspector must identify possible Underground Storage Tanks (USTs) and Aboveground Storage Tanks (ASTs) on the property and adjoining the property and note any vent pipes, fill pipes, concrete pads, saw cuts in paved areas, or other customary apparatus or indications of storage tanks.

13.1 Onsite Underground and Aboveground Storage Tanks

- ☐ Many UST and AST hazards have been previously defined and have existing corrective action plans, communications, or public correspondence on file at the state office responsible for storage tanks and leaking tanks.

- ☐ The environmental database report will indicate the presence, size, contents and number of registered tanks at a site.

- ☐ Leaking USTs (LUSTs) that have been reported will also be identified in the database report.

- ☐ The inspector must validate the database findings for the subject site and any adjacent sites.

- ☐ If a LUST site has been identified in close proximity and the inspector does not know whether the LUST is upgradient, downgradient, or crossgradient, he/she should verify this information. As a general rule, when actual subsurface information is nonexistent, the inspector may interpolate surface topography from USGS maps, and indicate this in the report.

13.2 Tank Information for ESA Report

☐ For each existing tank, the inspector should determine:

- Whether and when the tank has been registered with the appropriate regulatory agency
- Its size (capacity)
- Material of construction
- Age of material stored
- Corrosion protection method
- Containment facilities and leak protection equipment
- Current content status (e.g. empty, partially full)
- Results of any available leak tests and inventory reports, if available, and include them in the report

13.2.1 Removed or Abandoned Tanks

☐ If tanks have been removed or abandoned in place, also include available information describing the removal or abandonment procedure, and soil analytical data, as well as date of removal or abandonment.

☐ If possible, the inspector should determine, through interviews or reviews of regulatory records, the:

- Date when any removed or abandoned tanks were last operated,
- The name of the last owner and operator, and
- When the tanks were removed or otherwise abandoned or made inoperable.

☐ Include a map of all known locations of existing and former tanks in the report.

13.3 Offsite Leaking Underground Storage Tanks (LUSTs)

☐ The inspector must review state and federal lists of Leaking Underground Storage Tanks (LUSTs) within a ½ mile (0.8 km) radius of the site.

☐ The ESA report should include an assessment of the potential for environmental degradation at the site due to the LUST sites identified in the environmental database report.

13.4 Assessing Risks, LUST Sites

☐ As a general rule, the following six LUST site risk ratings have been utilized by the author extensively:

- High risk
- Moderate to high risk
- Moderate risk
- Low to moderate risk
- Low risk
- No risk

☐ Risk ratings for LUST sites are typically used in the following circumstances:

13.4.1 High Risk

- An onsite tank has a reported leak
- Petroleum soil contamination (not overspill) is present from an onsite tank
- An offsite upgradient tank (immediately adjacent or within ⅛ mile of the subject site) has a reported leak

13.4.2 Moderate to High Risk

- An onsite tank has a suspected leak

- An offsite tank with an unknown gradient (immediately adjacent or within ⅛ mile of the subject site) has a reported leak
- An offsite upgradient tank with a reported leak is within ¼ mile of the subject site
- Unregistered abandoned tanks are onsite
- Tanks are onsite but their status is unknown

13.4.3 Moderate Risk

- Onsite tanks have been removed
- An offsite upgradient LUST is over ¼ mile away from the subject site
- An offsite crossgradient LUST is immediately adjacent to the subject site

13.4.4 Low to Moderate Risk

- Onsite tanks have been removed or closed in place and no reported LUSTs are within ⅛ mile of the subject site
- No onsite tanks exist and no reported LUSTs are within ⅛ mile of the subject site

13.4.5 Low Risk

- No tanks are onsite and no reported LUSTs are within ¼ mile of the subject site

13.4.6 No Risk

- No tanks are onsite and no reported LUSTs are within ½ mile of the subject site

***

Geology and Soils

A site's soil and geologic character will assist the inspector in determining the susceptibility of groundwater to contamination. For example, aquifers or water-bearing zones located beneath impermeable layers of rock or soil are less susceptible to near-surface contaminant sources, such as leaking underground storage tanks.
Potential data sources include the USGS and state and local environmental agencies.

14.1 Site Geology

☐ For the site geology portion of the ESA, examine geologic data, to the extent that they are available, and if possible, the physiographic province within which the property is located, including but not limited to the following characteristics:

- Rock types
- Impermeable layers
- Bedrock characteristics of local geologic formations
- Seismological data if appropriate

14.2 Soils Investigation

☐ For the soils investigation, identify soil types and general characteristics, such as:

- Drainage
- Permeability
- Soil mantle depth
- Depth to bedrock

- Soil classifications

☐ Soil data assists in determining the:

- Ease with which contaminants may migrate laterally and downward
- Depth to groundwater
- Suitability for underground storage tanks

☐ Some soils are highly corrosive to metals and may contribute to the failure of steel tanks.

☐ Soil data can be obtained from county soil surveys. These surveys are published by the U.S. Department of Agriculture (USDA) Soil Conservation Service, and are available for most counties throughout the U.S.

***

Topography and Hydrology

In the absence of site-specific hydrogeologic data, surface topography can usually be used to infer the direction of groundwater flow (and, therefore, the direction of flow of contaminated waters) in near-surface aquifers.

15.1 Site Topography

☐ Examine geologic and hydrologic data, to the extent that it is readily available, focusing on the range of site elevations, overall site topography or slope, and significant physiographic features.

15.1.1 Physiographic Features

☐ Significant site physiographic features include:

- Ridges
- Streams
- Valleys

☐ The inspector should always examine hydrogeologic data to the extent that it is readily available.

☐ The U.S. Geological Survey (USGS) 7.5-minute topographic (Quadrangle) maps can be used for most areas of the United States.

15.2 Site Hydrology

☐ Hydrologic data is important in assessing the significance of water contamination. For example, contaminated groundwater within a naturally poor quality aquifer with no withdrawal for human use or consumption is typically considered insignificant.

☐ Regarding groundwater, the inspector should attempt to determine:

- Approximate depth to groundwater
- Aquifer productivity
- Current source of drinking water
- Direction of flow
- Nearest producing wells
- Suitability for domestic consumption
- Water quality

☐ The nearest producing wells should be identified as to location, pumping rate, and type of use.

15.2.1 Surface Water

☐ For surface water, the inspector should describe site drainage and water quality.

☐ Potential sources of hydrologic data include the USGS, and state and local environmental or water resource agencies.

15.2.2 Stormwater Drainage

☐ Based on federal, state, and local requirements that pertain to the subject property, determine and discuss the compliance requirements.

☐ The inspector should assess whether the property drainage system, its device, its design, and its operation are in compliance with any governmental requirements pertaining to stormwater permits.

***

Hazardous Substances

The inspector should always identify any locations used for the disposal of hazardous wastes and nonhazardous wastes on site maps and floor plans.

16.1 Hazardous Substances

☐ Review appropriate records for hazardous waste or hazardous substance activities at the site. This includes the identification of any:

- Existing site EPA identification (ID) number
- Permits
- Manifests
- MSDS sheets
- Methods used to dispose of solid waste or waste waters

16.2 Storage and Disposal Areas

☐ Observe the storage areas for waste materials to determine whether any adverse environmental conditions exist as a result of improper storage facilities or practices.

☐ Evaluate the site for evidence of onsite disposal or treatment of waste materials.

☐ In the summary and recommendations section of the report, include brief comments on whether any disposal and treatment methods seem appropriate.

- ☐ Also indicate whether such methods appear to meet applicable standards (if the information available supports such a conclusion).

- ☐ State any assumptions used in this conclusion.

16.3 Hazardous Chemicals

- ☐ Identify any hazardous or potentially hazardous chemicals or raw materials in connection with the site(s) or facility(ies) that are:

    - Used
    - Generated
    - Stored
    - Released
    - Transported
    - Disposed

***

Asbestos-Containing Materials (ACMs)

In 1973, EPA banned the use of sprayed-on or troweled-on friable materials. By 1979, sprayed-on Asbestos-Containing Materials (ACMs) were no longer allowed in building construction. ACMs have been used extensively in many schools, public buildings, and private residences. Asbestos surveys are not typically required by Phase I ESA SOWs; however, the information is provided here to familiarize the inspector with the subject matter.

17.1 Asbestos Potential

☐ The potential for Asbestos-Containing Materials (ACMs) should be evaluated for all buildings constructed prior to 1987 that have any remaining useful life, excluding single-family (one-to-four units) dwellings.

☐ Three forms of asbestos are typically found in buildings. They are:

- Surfacing materials
- Thermal system insulation
- Miscellaneous forms

17.2 Asbestos Survey

☐ In the event a full Comprehensive Asbestos Survey (CAS) is required by the SOW, the Comprehensive Asbestos Inspection must be conducted in accordance with the Asbestos Hazard Emergency Response Act (AHERA) and any state regulations in force at the time of the CAS.

- ☐ Comprehensive asbestos surveys and inspections require the inspector to:

  - Make visual inspections of the building units to determine the presence of suspect asbestos-containing building materials
  - Identify locations of homogeneous areas of suspect ACMs from which samples are to be obtained
  - Collect bulk samples of suspect ACMs
  - Provide selected photographs of the types of suspect materials sampled
  - Document on a building blueprint, or floor plan sketch, the locations of suspected ACMs
  - Estimate the total quantity of ACMs (calculated on a square footage basis for surfacing and miscellaneous materials, and on a linear footage basis for pipe lagging, duct tape, etc.)
  - Provide information on the friability of the ACM
  - Provide information on the overall condition of the material
  - Submit asbestos samples for testing and analysis by a NVLAP Certified Laboratory

- ☐ Upon completion of the Comprehensive Asbestos Survey (CAS), the inspector must prepare and present a written compilation of data to classify and identify ACMs, their condition, potential for disturbance, and hazard ranking.

***

Lead

If any structures on the subject site were constructed prior to 1978, the inspector should indicate whether the property is in compliance with state, local, and municipal laws, regulations, or ordinances regarding lead-based paint inspection, abatement, and notification requirements.

18.1 Lead-Based Paint Survey (LBPS)

- ☐ The inspector should indicate the age of all residential structures on the site. Residential structures include single-family dwellings (one-to-four units), multifamily dwellings, mixed-use residential dwellings, and mobile homes.

- ☐ If the property is located in a state or local jurisdiction that requires inspection and testing of lead-based paint, the inspector should notify the site representative of the requirements prior to performing, or subcontracting for, the necessary inspection and testing services.

- ☐ The process for conducting an LBPS and required records are similar to the CAS described in the previous section, and includes:

    - Documenting on a building floor plan the locations of suspected LBP
    - Estimating the total quantity of LBP, on a square footage basis
    - Providing information on the overall condition of the LBP

- ☐ All inspectors and subinspectors must state in their ESAs that they are properly licensed and/or certified to conduct inspections and testing for lead-based paint.

18.2 Lead in Drinking Water

- ☐ The inspector should indicate if the property is served by a publicly regulated municipal water service, and whether the local utility providing the drinking water meets current EPA standards for lead concentrations.

- ☐ The inspector should include references to any documented reports regarding lead in water impacting the subject property in the ESA study.

- ☐ The inspector should include data supporting assumptions pertaining to any recommendations to conduct Phase II sampling for lead in drinking water.

- ☐ The inspector should include supporting data in the summary and recommendations section of the report.

- ☐ The inspector should include, in the appendices, copies of the salient features of any reports obtained.

***

Polychlorinated Biphenyls (PCBs)

The inspector should identify the presence of PCB contamination based on observation and review of facility records regarding the site. If electrical or other equipment has the potential to contain PCBs, examine available records related to its use to evaluate compliance with state and federal regulations, and indicate whether equipment is owned by an electric utility.

19.1 Polychlorinated Biphenyls (PCBs) Identification

☐ If the owner of PCB-related equipment is identified, the inspector should contact the owner to determine the PCB content. The owner is typically the local Public Service or electric utility company.

☐ Potential PCB-containing equipment and items are easy to identify. Typical PCB-related equipment and items include but are not limited to:

- Capacitors
- Circuit breakers
- Cutting oils
- Heat transfer systems
- Hydraulic fluids
- Lamp ballasts
- Lubricants
- Plasticizer applications
- Transformers
- Vacuum pumps

19.2 What the Inspector Should Do — PCB Equipment Locations

☐ For each PCB-related equipment location, the inspector should:

- Assess whether PCBs may be present in some quantity
- Determine PCB content of oils or other such fluids
- Inspect for evidence of spills and releases
- Identify the name of the utility company (if applicable) or unit manufacturer
- Identify equipment serial numbers and other descriptive markings
- Identify the date of manufacture
- Assess conditions that may result in future harm
- Locate suspect equipment on site maps and facility building plans

19.3 PCB Disposal

☐ The inspector should also determine the onsite locations of disposal of equipment, oil, or other substances used or released in connection with the operation of suspect PCB equipment.

***

Landfills, Pits, Sumps, Drywells, and Catchbasins

The inspector should identify in the ESA study whether the site is located on or within a ½-mile radius of a known licensed or unlicensed landfill, or a federal/state-permitted hazardous waste/disposal site.

20.1 Landfills

- ☐ Landfills should be identified on the environmental database report.

- ☐ From the environmental database report, the inspector should be able to locate the landfill, and conduct a site visit if the landfill is within ½-mile of the subject site.

- ☐ In considering whether there is any probable resulting groundwater or soil contamination migrating toward the subject site, the inspector must consider:

    - Landfill distances from subject site
    - Gradient orientation in relation to subject site
    - Laboratory results of any soil and/or water sampling conducted on and off the landfill
    - Current regulatory status of the landfill

- ☐ After reviewing federal, state, and local environmental agency records and conducting site reconnaissance, the inspector should include answers to the following questions in the ESA:

    - Whether it is an open or closed landfill
    - Who the owner/operators are

- Whether any subsurface investigations have been conducted
- Whether there are any monitoring wells onsite
- Whether any sampling results have been reviewed by regulatory agencies
- Whether any corrective actions are being requested by reviewing agencies

20.2 Pits, Sumps, Drywells, and Catchbasins

☐ The inspector should report on the location, condition, and contents of all readily accessible:

- Pits
- Sumps
- Surface impoundments

☐ The inspector should note the current disposition of such pits, sumps, drywells, and catchbasins and surface impoundments in the report, and indicate their locations on a map of the site.

☐ Pits, sumps, drywells, and catchbasins and surface impoundments have to be assessed because of their propensity for waste collection, dumping activities, and their ability to interact with subsurface water.

☐ The inspector should also identify whether these items are registered with the appropriate regulatory authority.

***

Section 21

Radon

If the property is in an area of EPA-determined high concentration of radon, or if there is a state or local requirement to measure radon levels, the inspector should acknowledge this in the ESA.

21.1 Radon Assessment

☐ If the subject site has residential building(s) on it, the inspector should attempt to identify whether the site is located in an area of EPA-determined high concentration of radon. If so, the inspector should contact the client and request to have a test performed which meets EPA requirements for reliability-in-testing (e.g., Radon Measurement Proficiency Program).

☐ Upon completion of the radon survey, the inspector should provide appropriate recommendations in the hazards investigation and summary and recommendations sections of the ESA report.

21.1.1 Radon Certification

☐ As with testing for asbestos and lead-based paint, all inspectors and subinspectors must state in their report that they are properly licensed and/or certified to conduct radon tests and provide evidence of such licenses and/or certifications. If an inspector does not have the required certification, he/she should contact the appropriate agency and request a list of certified inspectors to use as subconsultants.

***

Section 22

Special Resources

Special resources are natural, cultural, scientific, or related issues that may pose positive or negative impacts to sites.

22.1 Special Resources Overview

☐ Based upon maps, aerial photographs, a title search, adjacent land ownership, and site reconnaissance, the inspector should ascertain whether the subject site is undeveloped and/or adjacent to or contiguous with any lands managed by a governmental agency (federal, state, or local) primarily for cultural or natural resources. Resources considered special include:

- Archaeological resources
- Historic properties, sites and districts
- Coastal zones
- Threatened and endangered species
- Wetlands/Floodplains
- Wild and scenic rivers
- Scientific significance
- Wilderness areas
- Natural landmarks
- Sole-source aquifers
- Open space or recreational usage
- Wildlife sanctuaries or refuges
- Other conservation purposes

22.2 Special Resource Information Requests

☐ When requesting special resource verifications from government agencies, the inspector should include:

- A copy of the pertinent USGS topographic map (7.5 Minute 1:24,000 scale) with property boundaries delineated on the map(s)
- The property's name and address, including city, county, and state
- The legal description, including lot, block, township, range and section (as applicable)
- A general description of the property's physical characteristics, including acreage (this information can be obtained from the property appraisals, the county or city assessor or planning departments, or a site visit)

***

Endangered Species

Endangered species are those species in danger of extinction throughout all or a significant portion of their native habitat or range. If the site is greater than 50 acres in size, the inspector should determine whether any endangered species and/or species' habitat exists on the property.

23.1 Endangered Species Identification

☐ The inspector should identify whether any critical or unique habitat exists on, or adjacent to, the subject site. This is defined as an essential segment of habitat that contains the unique combination of conditions necessary for the continued survival of an endangered species, such as:

- Soils
- Vegetation
- Predator species

☐ If critical or unique habitats are present and the site meets the size requirements outlined above (50 acres), the inspector needs to:

- Determine if the site is regulated
- Identify the types of activities regulated
- Identify the regulatory agency

☐ The U.S. Fish and Wildlife Service or a state fish and wildlife agency is typically the responsible entity for endangered species issues.

***

Historic Properties and National Landmarks

Historic properties include any sites, structures, building objects, improvements, properties within historic districts, or properties with archaeological resources that are either listed or eligible for listing in the National Register of Historic Places.

24.1 Historic Property Identification

- ☐ Based upon maps, aerial photographs, site reconnaissance, and other information on the property, the inspector should determine if there are any:

  - Improvements on the site 50 years of age or older

- ☐ If the age requirement is met, the next step is to determine if there are any:

  - Undeveloped areas on the site greater than one acre in size

- ☐ If the age and size requirements are met, the inspector should contact the State Historic Preservation Officer (SHPO) and provide the SHPO with information on the property so that the SHPO can determine whether any regulated historic resources exist at the site.

24.2 Historic Site Research Objectives

- ☐ In determining the presence of cultural and historic resources, phone contact to the SHPO should be made first to determine what information is required. The inspector's objectives are to:

  - Determine whether the historic property is regulated

- Identify the types of activities regulated and the responsible regulatory agencies

24.3 Historic Site Documentation

☐ Typical information requested by the SHPO may include:

- Photographs of the site and its improvements (including photos of all sides of any buildings or other structures)
- The property's name and address, inclusive of city, county and state
- The general description of the property's physical characteristics, including acreage, type of property (land, commercial, residential, multifamily, etc.)
- The number of building structures on the site, building size (square feet), and number of stories in building
- The year that structures were built (age)
- A description of the present condition of the structure

☐ For properties with an acre or more of natural area, contact may also have to be made with the Advisory Council on Historic Preservation (ACHP). The SHPO should be able to advise the inspector if this step is necessary. If so, the inspector may have to submit a USGS map and an aerial photograph with the property boundaries delineated on it.

☐ Additional maps such as a tax map, surveyors plat or map, floodplain map, or a neighborhood/site map may also be requested.

24.4 National Landmarks

☐ Based upon the National Registry of Natural Landmarks, maps, aerial photographs, and site reconnaissance, the inspector should determine whether there are any national natural landmarks on the site.

***

Recreational Areas

Recreational areas and open space are key environmental considerations that may pose positive or negative impacts on a site. The major issue is primarily in terms of existing or potential public recreational value.

25.1 Recreational Area Identification

☐ Based upon maps, aerial photographs, and site reconnaissance, the inspector should determine whether there are any recreational areas, including areas that are appropriate for use as a park or other outdoor recreational activity.

☐ If so, contact should be made with the appropriate:

- State recreational authorities, including the State Outdoor Recreational Planner, or the person with those responsibilities
- Local recreational authorities

☐ The primary purpose of these contacts is to determine whether the site has been identified for acquisition, or set aside for recreational purposes.

☐ The existence of recreational areas on or adjacent to a site may limit development opportunities, but may increase the value of the site.

***

Scientific Significance

Scientific significance in a site may enhance property values, and tends to attract the public and special interest groups interested in conserving and/or researching the resource.

26.1 Identifying Scientific Significance

☐ Based upon maps, aerial photographs and site reconnaissance, ascertain whether there are any areas with scientific significance, such as:

- Paleontological resources
- Artifacts
- Fossils
- Historic properties valued for scientific study or research
- Areas set aside for scientific research

☐ Sites with known Native American artifacts, or cultural and anthropological resources, are subjected to strict scrutiny by governmental and quasi-governmental entities.

26.2 Agency Responsibility

☐ The federal agency typically responsible for scientific values issues is the Advisory Council on Historic Preservation (ACHP).

***

Sole-Source Aquifers

Sole-source aquifers are the sole or principal source of drinking water, as established under Section 1424(e) of the Safe Drinking Water Act, and are special resources which, if contaminated, would create a significant hazard to public health.

27.1 Sole-Source Aquifer Identification

☐ Based upon the EPA list of sole-source aquifers and upon site reconnaissance, the inspector should ascertain whether the site is located within the boundaries of a sole-source aquifer that has been designated or proposed for designation by the EPA under the Safe Drinking Water Act. In assessing this resource, the inspector should determine whether the site is undeveloped and has features that would contribute to the recharge of the aquifer, such as:

- Wetlands
- Watercourses
- Springs
- Sink holes
- Karst topography

☐ The inspector should determine if financial or other assistance provided through governmental programs for site improvement would be regulated, and the types of assistance restricted. The inspector should also identify the responsible regulatory agency (i.e., EPA or the appropriate state office of groundwater protection).

***

Coastal Zones

This special resource consists of undeveloped coastal dunes and beaches that are protected and preserved within the scope of the Coastal Zone Management Act.

28.1 Undeveloped Coastal Zones

☐ Based upon maps, aerial photographs, and site reconnaissance, the inspector should ascertain whether there are any undeveloped coastal areas that fall within the scope of the Coastal Zone Management Act, including:

- Determining whether development of the site is regulated
- Identifying the types of restrictions
- Identifying the responsible regulatory agency

☐ The responsible agencies are usually the state Coastal Zone Management Agency and the local unit of government with jurisdiction for the area.

☐ Development restrictions typically in force in these areas relate primarily to building infrastructure and foundation development, height and aesthetic restrictions, and other zoning or use restrictions.

☐ When requesting information on these areas, site documentation (outlined in Section 22.2) is typically required if a formal response is needed.

***

Wetlands

Wetlands are typically low-lying lands which are near bodies of water, periodically covered by fresh, brackish, or salt water, and largely covered by vegetation. These areas are saturated by surface or groundwater at a frequency and duration sufficient to support the growth of hydrophytic vegetation typically adapted for life in saturated soil conditions.

29.1 Wetlands Identification

☐ If the site is greater than 50 acres in size, the inspector should obtain wetlands maps, National Wetland Inventory maps, or Soil Conservation Service wetland determination maps from federal, state, or local agencies.

☐ The primary contacts for wetlands issues are the Army Corps of Engineers (ACOE), the EPA, and the USFWS. First contact should be made with the local ACOE office. Based on maps, aerial photographs, and upon site reconnaissance, the inspector can ascertain whether there are wetlands or restorable wetlands (as defined in the Clean Water Act, as amended, or the Emergency Wetlands Resources Act of 1986) on the site. Wetlands are usually associated with:

- Swamps
- Marshes
- Bogs
- Intermittent creeks
- Streams
- Lakes

☐ Wetlands are subject to strict development restrictions under the Clean Water Act.

29.2 Wetlands Review Process

☐ The inspector should incorporate the following information in the wetland identification process:

- The presence of non-hydric soil that exhibits hydric characteristics
- The presence of hydrophytic vegetation
- The presence of a wetland hydrologic regime
- Whether the wetland is regulated
- The types of activities regulated
- Whether wetland restoration is required as a condition of a governmental permit or order

☐ The responsible regulatory agencies for wetland issues include:

- Army Corps of Engineers (ACOE)
- Environmental Protection Agency (EPA)
- National Park Service (NPS)
- U.S. Fish and Wildlife Service (USFWS)
- Soil Conservation Service (SCS)

29.3 Wetlands Delineation

☐ Wetlands delineation is a specialized field that requires indepth knowledge of biological, horticultural, hydrological, and related issues. The Phase I inspector should not attempt to conduct any formal wetlands delineation unless he/she is trained and knowledgeable in the field.

☐ The ACOE may be requested to delineate a wetland, or they may direct the inspector in the conduct of a certified delineation.

***

Floodplains

Undeveloped floodplains are addressed in Executive Order 11988, "Floodplain Management," which defines 100-year floodplains. Many property-use restrictions exist in floodplain areas.

30.1 Undeveloped Floodplains

☐ Based on Flood Insurance Rate Maps (obtained from the Federal Emergency Management Agency), a copy of the site appraisal, aerial photographs, and upon site reconnaissance, the inspector should ascertain whether there are any undeveloped floodplains on or adjacent to the site.

☐ Development restrictions are typical in floodplain areas. Contacting the local government responsible for them is usually the best means of obtaining specific information.

☐ If the site is situated in a 100-year floodplain, the first thing the inspector must do is determine if the area is regulated.

☐ Next, the inspector should identify the types of activities restricted. Then, he/she should identify the regulatory agency (responsible local government entity).

***

Section 31

Wild and Scenic Rivers

A Wild and Scenic River is a river and the adjacent area within the boundaries of a component of the Natural Wild and Scenic Rivers system.

31.1 Wild and Scenic River Identification

☐ The inspector should ascertain whether there are any wild and scenic rivers that are either designated or proposed for designation under the Wild and Scenic Rivers Act on or adjacent to the site, based on: a copy of the site appraisal, a list of wild and scenic rivers, aerial photographs, and site reconnaissance.

☐ If a listed wild and scenic river does exist, the inspector should:

- Identify the agency responsible for administering the wild and scenic river area, if any
- Identify the name of the area and recommend an agency contact
- Attempt to determine what, if any, impacts are posed by the wild and scenic river on the subject site

☐ If the inspector cannot determine what the potential wild and scenic river impacts are, a copy of the agency contact data (included in the appendices of the ESA report) with a recommendation to the client to make the contact with the agency staff interviewed should meet the Phase I SOW.

***

Wilderness Areas

Wilderness areas are undeveloped federal lands retaining their primeval character and influence, without permanent improvements or human habitation, that is primarily protected and managed for preservation in its natural state. If the site is greater than 50 acres in size, the wilderness area issue may be pertinent.

32.1 Wilderness Area Identification

☐ The inspector should identify whether the site is contiguous with, or adjacent to, federally owned lands or lands legally designated for acquisition by a federal agency, that are designated or proposed for designation as a part of the National Wilderness Preservation System (NWPS).

☐ Information concerning NWPS issues can be obtained from:

- Available special resource lists
- Reviews of wilderness area maps
- Aerial photographs
- Site reconnaissance

☐ For wilderness area issues, the inspector should:

- Identify the agency responsible for administering the area, typically the National Park Service (NPS), Forest Service (FS), and/or U.S. Fish and Wildlife Service (USFWS)
- Identify the name of the area and recommend an agency contact
- Attempt to determine what (if any) impacts are posed by the wilderness area on the subject site

- ☐ If the inspector cannot determine what the potential wilderness area impacts are, a copy of the agency contact data (included in the appendices of the ESA report) with a recommendation to the client to make the contact with the agency staff interviewed should meet the Phase I SOW.

- ☐ When making agency contacts by mail or in person, the data outlined in Section 22.2 is typically required.

***

Site Maps

> *A sequence of maps at various scales is typically required to prepare a Phase I ESA.*

33.1 Primary Maps

- ☐ The inspector should always start with a 7.5 Minute USGS Quad Map for a large area, small scale view of the subject site's area topography.

- ☐ Next, it is advisable for the inspector to use a city map, Sanborne map, and/or fire insurance maps (if available) for a medium scale local area view. Then the inspector should obtain, or prepare, a site map or plat for a larger scale site specific overview.

- ☐ If structures are onsite, the inspector should utilize actual floor plans (if available), and if not draw floor plans of the structure as close to scale as possible.

33.1.1 Map and Site Plan Sources

The primary sources for ESA maps and plans are as follows:

- ☐ 7.5 Minute USGS Quad Map (check local phone directory to see whether a USGS sales counter is located in your area). The national USGS map sales mailing address is:

USGS Map Sales
P.O. Box 25286
Denver, CO 80225

- ☐ As of this writing, standard 7.5 Minute Quadrangles cost $2.50, plus a $1.00 handling charge. Order forms that list all map products and prices are available from USGS.

- ☐ City maps — county and local government planning, zoning, and/or engineering departments

- ☐ Site maps or plats — county tax assessors and local government planning, zoning, building, and/or engineering departments

- ☐ Actual floor plans — the owner/client, onsite property manager or local government building department

33.2 Secondary Maps

- ☐ If special resources are on or near the site, various types of maps should be available either individually, or included in studies, such as:

    - Floodplain maps
    - District maps
    - Conservation area maps
    - Seismologic maps
    - Zoning maps
    - Comprehensive plans
    - Wetlands maps, etc.

***

Sample Phase I ESA

The ESA Cover Letter...

- ☑ Is preferable but not mandatory
- ☑ Provides client and site identification
- ☑ Gives status of Phase I ESA tasks
- ☑ Summarizes Phase I ESA findings
- ☑ Provides Phase I ESA recommendations
- ☑ Is a convenient place to attach an invoice
- ☑ Provides contact person and phone number

September 17, 1995

Mr. John Doe
JDE Enterprises
4000 Main Street, Suite 300
Denver, Colorado 80112

RE: Phase I Environmental Site Assessment
JDE Oil - Asset No: 656310912
Brentwood, Colorado

Dear John:

Coopers Environmental Information, L.L.C. (CEI) has recently completed a Phase I ESA on the aforementioned property. All tasks are now complete and the findings and recommendations are as follows:

- After complete sampling and analytical tasks were conducted, no asbestos-containing building materials were identified onsite.
- The interior has been cleaned out since our first investigation, and only small containers (5-gallon and less) utilized for paints, solvents, and cleaning materials were identified on the day of our investigation.

Recommendations

- CEI recommends the empty 55-gallon drums, bulk storage containers, non-operational ASTs, and miscellaneous debris be removed from the site and disposed of properly.
- As some of the drums, containers, and ASTs identified may have contained hazardous substances, CEI recommends that any item still containing fluids be sampled prior to disposal to determine whether it must be disposed of in accordance with hazardous materials regulations.

If you have any questions or comments regarding this project, please feel free to call us at 555-6323 at your earliest convenience.

Sincerely,
Coopers Environmental Information

The cover letter is basically a reiteration of the summary (findings) and recommendations sections of the ESA!

The ESA Title Page identifies...

- ☑ Type of inspection
- ☑ Site location and any identifying numbers used by client
- ☑ Client name and address
- ☑ Inspector name and contact information
- ☑ Date of ESA
- ☑ Final or Draft
- ☑ Inspector name and signature

PHASE I ENVIRONMENTAL SITE ASSESSMENT

JDE Oil Site
124 E. Arapahoe Avenue
Brentwood, Colorado
Asset No: 656310912

prepared for

JDE Enterprises
4000 Main Street, Suite 300
Denver, Colorado 80112
Attn:
Mr. John Doe

prepared by

Coopers Environmental Information
7600 E. Arapahoe Road, Suite 114
Englewood, Colorado 80112
PH: (303) 555-6323/FAX: (303) 555-6394

Date: September 17, 1995

Submitted by

____

André R. Cooper, R.E.A.
President and CEO

Adopt a standard cover sheet that can be used as a boilerplate. In actuality the cover serves as an 8.5" x 11" business card!

The ESA Table of Contents...

☑ Is based on the primary sections of the project SOW

☑ Is broken down in order, providing a logical flow of tasks conducted

☑ Provides a list of appendices

☑ Is a boilerplate that can be reused whenever the same project SOW is in effect

Table of Contents

This table of contents was created automatically in Wordperfect 5.2 using the mark text, TOC definition, and generation functions.

The Executive Summary...

- ☑ Identifies the site
- ☑ Provides inspection time and date
- ☑ Gives a site overview and facility description
- ☑ Provides occupancy status
- ☑ Provides brief description of surrounding area
- ☑ Reiterates the Hazard and Special Resource summaries

Phase I Environmental Site Assessment

JDE OIL
124 E. Arapahoe Avenue
Brentwood, Colorado

EXECUTIVE SUMMARY

A Phase I Environmental Site Assessment (ESA) was completed by Coopers Environmental, Inc. (CEI) for the petroleum distribution site located at 124 E. Arapahoe Avenue in Brentwood, Arapahoe County, Colorado. This work was performed for John Doe Enterprises, Inc. (JDE), Denver office. This Phase I ESA was conducted in accordance with the scope of work provided with the contract for environmental assessments. Investigation activities included agency interviews, site inspections, property transaction reviews, and governmental records reviews.

The Executive Summary should always be presented first, but written last!

Site reconnaissance activities began at approximately 11:00 a.m. on the 3rd of September, 1995. The structure is a relatively small prefab, metal, garage type structure. CEI could not find data to confirm the exact date of construction as there was no Municipal Building Department; however, as the County Assessor's records indicate that JDE had purchased all lots on 1/17/79, it is conceivable that the structure was probably erected in 1979 or 1980.

The subject site is square-shaped and comprised of five lots totalling approximately 16,250 square feet (125' x 130'). A prefabricated, rectangular, metal building (approximately 3,000 square feet) is on the western side of the property along Arapahoe Avenue.

On the day of our investigation the premises appeared to be occupied, as evidenced by interior lights on in the office area and a vehicle parked out front. The door had a handwritten sign indicating that the business had moved its operations on 9/2/95 to Denver, one day prior to our arrival.

The surrounding area is primarily agricultural, low-density residential, with some small commercial establishments along Arapahoe. No industrial facilities were identified in the immediate site area. Several empty and near empty ASTs, bulk storage containers, and 55-gallon drums were identified onsite (see Figure 2 in Maps Appendix).

3

The Executive Summary also describes...

- ☑ Preinspection tasks
- ☑ Sampling, if any
- ☑ Status of onsite reconnaissance visits
- ☑ Inspection times and dates of initial and subsequent visits, if any
- ☑ Other data of importance relating to ESA findings

Prior to visiting the site, our research indicated the rural area near Brentwood, Colorado had no Fire, Building or Municipal offices. Upon arrival, CEI visited the local office of the Bureau of Land Management (two blocks west of the subject site) to conduct interviews and collect historic data on the site and surrounding area. BLM was the only agency with a presence in Brentwood.

After complete sampling and analytical tasks were conducted, no asbestos-containing building materials were identified onsite.

On our second visit to the site, (September 27, 1995), the interior had been cleaned out (see photos), and only small containers (5-gallon and less) utilized for paints, solvents, and cleaning materials were identified on the day of our investigation.

Summarize results and findings, as well as the availability or unavailability of information sources!

4

The Scope of Services (SOS)...

- ☑ Reiterates Statement of Work (SOW)
- ☑ Helps familiarize persons unaware of specific contract details with the scope and nature of the project
- ☑ Forms project plan
- ☑ Is a boilerplate based on SOW
- ☑ States what was supposed to be done
- ☑ States what was actually done, considering factors such as site access, building access, tenant cooperation with the inspector, etc.

1.0 Scope of Services

CEI performed a Phase I ESA, including photographic inspection of the site and adjacent properties. Phase I site reconnaissance tasks were completed by Mr. Andre Cooper, C.E.P., on June 3, 1995. Historical records and regulatory database searches were conducted to obtain information on past land use activities that may have been environmentally significant.

The Phase I ESA services provided were in accordance with the CEI's 1995 Scope of Service. Services provided included:

Site History

To conduct tasks related to reconstruction of the site's history, CEI:

- researched available deed and title information to identify owners who may have used the property for industrial and/or hazardous waste handling activities.
- obtained and reviewed aerial photographs of the site and adjacent properties to determine current uses that may have an environmental impact on the site.
- obtained and reviewed available governmental records to identify use, generation, storage, treatment, and/or disposal of hazardous materials, and release incidents which may impact or have impacted the site.
- reviewed available reports and other documentation from government agencies on the site and adjacent lands.
- interviewed site area representatives to identify any additional areas of concern.

To avoid SOS problems with the client, always reiterate the Statement of Work in the ESA!

Environmental Database Searches

To conduct tasks related to environmental database searches, CEI identified, obtained, and reviewed federal, state, and local databases and records to discover if the site, or any adjacent sites, have posed an environmental hazard in the past, or currently pose such a threat.

An environmental database report is included in the appendices. Database and information researched included the following:

5

The Scope of Services outlines data searches of...

- ☑ **Federal records and agencies**
- ☑ **State records and agencies**
- ☑ **Local records and agencies**
- ☑ **And describes any unavailable records, for the site area**

Federal Records Searched

NATIONAL PRIORITIES LIST (NPL) - The U.S. Environmental Protection Agency (EPA) listing of federal Superfund sites. The NPL is a list of sites that have been evaluated using the Hazard Ranking System, and that have been determined to pose an imminent threat to human health or the environment.

CERCLIS - EPA database of known, alleged, or potential hazardous waste sites that have been investigated, or require investigation, under CERCLA. For sites that have been investigated, site status is given; sites not investigated are potentially eligible for inclusion on the NPL.

RCRIS (Resource Conservation and Recovery Information System) - Database of sites that generate, store, transport, treat, and/or dispose of hazardous wastes. Sites that fall into these categories are required to submit a notification of hazardous waste activity to EPA. RCRA facilities are classified as: Large or Small Quantity waste generators.

RCRA - Treatment, Storage and Disposal Sites (TSD). Companies which have reported that they treat, store, and/or dispose of hazardous waste; and RCRA Transport Sites: companies which have reported that they transport hazardous wastes.

ERNS - The Emergency Response Notification System (ERNS). The EPA database of reported hazardous substance releases.

State Records Searched

The following state records were reviewed for this audit.

Hazardous Waste Sites and Solid Waste Disposal Sites - Database for the identification and location of current and historical sites that engage in solid waste collection, transport, separation, recovery, and/or disposal activities.

Records Not Searched

The site area is considered rural/agricultural. During our database search effort, we discovered that the following state environmental database was unavailable, or not maintained, for the Brentwood, Colorado area: Underground Storage Tank (UST)/Leaking Underground Storage Tank (LUST) Records. According to our database search, the State of Colorado does not maintain a database for identification and locations of sites that maintain underground tanks that contain regulated (usually petroleum-related) substances in the site area.

6

State database names vary from state to state, but they are fairly consistent in content!

The Scope of Services also describes tasks involved with...

- ☑ **Site reconnaissance**
- ☑ **Visual inspections**
- ☑ **Exterior site analysis**
- ☑ **Interior inspections**
- ☑ **Special Resource investigations**

Site Reconnaissance

A reconnaissance visit to the site was made to inspect physical facilities, identify site drainage patterns, and make other observations. The Project Manager for this site investigation was André Cooper, R.E.A. The following tasks were conducted:

- Visual inspection of the site and surrounding properties was conducted. The inspection was conducted to identify potential USTs, ASTs, potential sources of ACMs, PCBs, or other hazardous chemicals.

- Exterior site analysis and photographs to identify potential hazards from onsite and offsite property uses.

- Visual surface investigations to reveal evidence of spills (stained soils or concrete) and vegetative stress.

- Interior site analysis and photographs to identify potential hazards from onsite operations.

Special and Cultural Resources

Visual observations and records investigations for special and cultural resources were conducted in the following areas:

Air Quality
Biotic and Wildlife Habitats
Unique Farmlands
Land Use
Socioeconomic Impacts
Recreation Areas
Wildlife
Archaeological Resources
Coastal Zones
Historic Resources
Threatened/Endangered Species
Transportation
Wetlands
Wild and Scenic Rivers

7

Even though it may seem redundant, it is a good procedure to state exactly what is to be done, prior to doing it!

Property/Asset Information portion lists...

- ☑ Property name
- ☑ Property address
- ☑ Property ID number
- ☑ Property manager
- ☑ Client
- ☑ Legal description

2.0 Asset Information

The pertinent asset information is:

Property Name:	JDE Oil Site
Property Address:	124 E. Arapahoe Avenue Brentwood, Arapahoe County Colorado 80112
Client Asset ID No:	656310912
Site Contact/Mgr:	John Doe
Onsite Property Mgr:	none
Client/Requester:	John Doe Enterprises, Inc. 4000 Main Street, Suite 300 Denver, Colorado 80112

The legal description for 124 E. Arapahoe Avenue is:

Lots 4, 5, 6, 7, and 8, Block 21 of the Original Town of Brentwood, Arapahoe County, Colorado.

8

Much of the property and asset information is obtained from information provided by the client. Always ask the client for this data!

Site Location and Description provides information on...

- ☑ Site usage
- ☑ Site vicinity
- ☑ Land uses
- ☑ Lot sizes/acreage
- ☑ Existing hazards
- ☑ Photos

3.0 Site Location and Description

The site is the location for JDE Oil's bulk sales and distribution operations. The property is located at 124 E. Arapahoe Avenue, in the City of Brentwood, Arapahoe County, Colorado. Brentwood is a small farming town located in northwestern Colorado with a population of about 800. The dominant land uses are agricultural: livestock (sheep/cattle/hogs), cattle feed, and sugar beet crops. Development in the area is extremely sparse, with a small mix of office, low-to-medium-density residential, and retail complexes along Main Street. No industrial facilities were identified in the immediate area.

During the site visit, a visual survey of the building and property was conducted for the presence of conditions which could indicate environmental impacts (e.g., open containers, tanks). Each lot is 25' by 130'. All five lots total 125' by 130', as indicated on the assessor's map (see appendices), and total square footage is 16,250 square feet. Per photographic documentation included in the ESA, the entire site is not developed.

As the site is used for the storage and distribution of petroleum-related products to local farmers, several Aboveground Storage Tanks (ASTs), bulk storage containers, and 55-gallon drums were identified onsite (see Figure 2 in Maps Appendix).

That ASTs that had fill gauges on them, indicated they were empty. On the day of our investigation it was noted that most of the ASTs had either been drained and were entirely empty or had minimal contents. Most of the 55-gallon drums noticed were unmarked and their contents were unknown. Some partially-filled drums and containers were identified onsite. Several ASTs were lying around the back and side yards, apparently abandoned (see photos in Appendix B). No signs of past or present leaks were observed around the ASTs or containers. A pile of old tires was noticed in the side yard, as was miscellaneous debris from past operations.

Three large (appx. 800-1000 gallon) plastic bulk containers used for motor oil, hydraulic and transmission fluids were tied to a tree in the northeast corner of the site. It appeared as though these containers had little or no fluids remaining (see photos in Appendix B).

9

Much of the site location and description information is obtained from real estate appraisals and site visits. Always ask the client for a copy of the most recent appraisal conducted on the subject site, it can save you a lot of time!

The Adjacent Property Descriptions include...

- ☑ Information on all sites immediately adjacent to the subject site, such as:
- ☑ Land uses and zoning
- ☑ Acreage
- ☑ Existing hazards
- ☑ Photos of all properties immediately North, South, East, and West of the subject site

Adjacent Properties

Dilapidated structures abut the site on lots to the north and south.

East of the site lies a 20-foot alley right-of-way with barns and open agricultural fields beyond.

On the west across Arapahoe Avenue lies a senior-citizen home, with small single-family residences beyond.

Toward the north along Arapahoe Avenue, also known as Main Street, small office and retail-type establishments were noticed.

It appeared as though the tallest building in town was a two-story structure.

No major industrial facilities are located in Brentwood, and none were identified on the day of our site reconnaissance.

10

The initial adjacent property description is a quantitative versus qualitative one. It is just a brief description of what exists on all sides of the subject site; potential hazards are discussed in the hazard review section!

The Site History section includes...

- ☑ A description of local or county history
- ☑ A description of area history
- ☑ A description of land uses
- ☑ An overview of the built environment
- ☑ An overview of the natural environment
- ☑ A description of readily available records

4.0 Site History

Arapahoe County was established in 1883 by an act of the Colorado territory legislature. Denver was established as the county seat in 1895. U.S. Highway 70 is the main east-west highway in the county. It intersects U.S. Highway 25, the main north-south highway, at Denver. State Highway 225 serves the southwestern part of the county, and State Highway 470 extends north and south through the town of Brentwood.

A system of all-weather secondary roads connects all parts of the county with the population centers of Denver and Brentwood. Access to these roads from ranches in the sparsely populated parts of the county is by winding dirt roads and trails. Many areas of the county are in native grass and are used for rangeland. Areas in the irrigated parts of the county commonly are used for irrigated pasture, with some farming.

As this site is in a rural portion of Colorado there were no records available indicating when the site was first developed. On the day of our investigation it appeared as though the City of Brentwood does not have a building department and does not maintain building permits. Based on the adjacent structures, it would appear as though a small wood frame building was onsite prior to the metal prefabricated structure. According to ownership records, the Doe's initially took possession of the site in 1977 (see ownership records in the appendices). As they own a bulk storage facility about a mile away from the subject site, distribution operations onsite may have started in 1977. Former owners are listed in Table 1, Section 4 (Site History) and title documents for 50+ years are included in the appendices. No Sanborn Maps or other Business Directories were available on the day of our investigation.

Wheat and other grains are marketed at elevators in Brentwood. Brentwood also has a large wool warehouse, which handles more than 3 million pounds of wool annually. The current operational status of this facility is unknown and it was not identified in the proximity of the subject site. Brentwood is headquarters for an Irrigation District and a rural electric cooperative. The Bureau of Land Management (BLM) also has offices in Brentwood.

Mr. Robert Underwood, P.E., Brentwood office of the BLM, was interviewed in person. The Brentwood BLM office is located 2 blocks from the subject site and Mr. Underwood indicated that he has lived in Brentwood over thirty years.

Researching site history is a major portion of the ESA! The primary goal is to identify whether past site activities have the potential of posing adverse environmental impacts.

11

The Site History also includes descriptions of...

- ☑ Interviews
- ☑ Site visits
- ☑ Topographic reviews
- ☑ Property transaction reviews
- ☑ Aerial photograph reviews
- ☑ Site operations

Mr. Underwood indicated that when he was a farmer in the 1960s and 1970s he used to purchase gasoline (diesel and regular) from the subject site. Mr. Underwood indicated that he was unaware of any environmental hazards at the site.

Conversations with the City of Brentwood's Water Department personnel revealed no environmental issues of concern. The same was true of CEI's discussions with Arapahoe County personnel. There were no onsite or adjacent site owners/operators available on the day of our investigation. According to conversations at the BLM, and phone calls to directory service, no local environmental, municipal, or fire officials were listed in the City of Brentwood. The results of these interviews are included throughout this study.

Based upon these interviews, there was no indication of fires or emergency response activities occurring at the site. There were also no indications of solid waste disposal activities occurring at the site. Environmental liens or government notifications relating to violations of environmental laws with respect to the site or abutting properties were not identified.

Based upon review of topographic maps USGS 7.5 Minute series Brentwood, Colorado quadrangle, property transaction records, and aerial photographs, the usage of the site has been primarily for the bulk storage and wholesale/retail distribution of petroleum products (regular/unleaded gasoline, diesel fuel, tractor/motor oil, transmission fluid, etc.).

An historic aerial photo for the early 1960s was available at the BLM (see appendices). The Brentwood City offices were staffed by one individual who indicated that he had lived in Brentwood for some time and had never heard of any environmentally-related problems at the site. City staff informed CEI that all engineering and building-related records were maintained at the county seat some 60 miles from Brentwood. Upon arriving at the Arapahoe County offices, no additional aerial photos were available. The 1980 Aerial Photo was the only readily available aerial of the subject site and it shows the site as it was on the day of our investigation.

CEI did not locate evidence of current or prior releases of petroleum or hazardous materials/wastes on the property. Evidence of prior site usage with respect to agricultural operations involving pesticides/herbicides is probable due to the nature of the area, but as Arapahoe is the main street, these types of onsite activities probably ceased decades ago (1940s).

12

When describing interview results use the interviewees' responses verbatim, when possible. Contact information for persons interviewed is included in the appendices!

The Site History portion of the ESA should also include detailed descriptions of information obtained on...

- ☑ Agency interviews
- ☑ Environmental permits
- ☑ Regulatory monitoring/reporting requirements (if any)
- ☑ Hazardous onsite operations
- ☑ The 50-year Chain-of-Title
- ☑ The previous owners/operators

No environmental permits were located for the subject site. It may be possible that because all tanks are aboveground they do not fall within the strict regulations, permitting, and reporting requirements that pertain to underground tanks. During our investigation it appeared as though the State was not concerned about aboveground tanks in farmland areas. No SPCC Plans were located during our investigation, it is possible that none exist. Brentwood is a small farm town with a main street that extends for 2 blocks. According to our findings there was no City Hall, no City Clerk, and no Fire Department. The best available information was obtained from the Bureau of Land Management (BLM).

Property transaction records, in the form of assessor's records, were reviewed at the Arapahoe County Tax Assessor and recorders offices in Denver, Colorado on February 3, 1995. A summary 50-year Chain-of-Title for this property is presented on the following page. Potentially hazardous operations unrelated to the bulk storage and distribution of petroleum related products were not identified on the day of our investigation. Owners prior to the JDEs appear to have been governmental or private individuals.

Table 1 - Property Transaction Records*

Date	Owner	Lot(s)
09/21/94	Capital Savings and Loan	4 thru 8
01/17/79	John and Sandi Doe	4 thru 8
05/23/77	John and Sandi Doe	5 thru 8
12/20/71	Arapahoe County Treasurer	4
12/21/70	Arapahoe County Treasurer	5 thru 8
10/06/65	Thomas DeGarcia	4
05/28/62	Arapahoe County Treasurer	4
12/16/43	Ed Wolden	5 thru 8
09/28/48	Rick and Florence Byrd	4
04/23/35	Richard Davidson	4
As of 1935	Joseph McIvey	4

*Copies of the above ownership records and mortgage documents are included in the appendices.

13

A brief summary of ownership records reviewed at a County Assessor's office is typically sufficient for the chain-of-title review. Copies of title documents should be included in the appendices!

The Regulatory Review typically consists of the following databases...

- ☑ NPL — Superfund waste sites
- ☑ CERCLIS — CERCLA sites or potential Superfund sites
- ☑ RCRIS — generators, treaters, storers, disposers
- ☑ RCRA — transporters
- ☑ ERNS — spills
- ☑ SWDS/SWLF — landfills
- ☑ USTs/ASTs — tanks
- ☑ A brief indication of unavailable records

5.0 Regulatory Review

Government records were reviewed by Coopers Environmental Information, on June 19, 1995. This review included Spill and Incident Reports, Resource Conservation and Recovery Act (RCRA) generator and handlers files, RCRA sites files, and Comprehensive Environmental Recovery, Compensation, and Liability Information System (CERCLIS) sites, as well as hazardous waste sites and solid waste disposal sites. The regulatory database search is included in the appendices.

Based upon these file reviews, the following information was noted:

The site and immediately abutting properties were not included on any regulatory database listing.

No RCRA TSD facilities were identified within 1 mile of the site.

The site and immediately abutting properties were not listed as RCRA hazardous waste generators/handlers. No small or very small quantity generators were identified within approximately .25 mile of the site.

No RCRA Large Quantity (hazardous waste) generators were identified within .25 mile of the site.

No Superfund (or potential Superfund) sites were identified within 1.0 mile of the site.

No National Priority List (NPL) sites were identified within 1.0 mile of the property.

No ERNS sites were identified in the site area.

No Landfills (SWLF) were identified within .75 mile of the property.

The environmental database report in the Appendices did not locate a UST or LUST database for this portion of the State of Colorado.

As this area (Brentwood, Colorado) is extremely rural and agricultural in nature, it is our opinion that Aboveground Storage Tanks are possibly the only types in use, in the immediate vicinity.

14

When conducting database searches, complying with minimum search distances is imperative!

The Site Investigation and Review of Hazards presents...

- ☑ Findings, data references, and interview results, included for each area reviewed

- ☑ Primary data sources for physical site environs including: USGS maps, soils surveys, SCS (Soil Conservation Service) floodplain maps

- ☑ Review of the above sources will help the inspector define the following site issues: topography, geology, soils, hydrology, and floodplains

6.0 Site Investigation and Review of Hazards

Topography

According to the topographic map prepared by the U.S. Geological Survey (USGS) 7.5 minute series Brentwood, Colorado quadrangle, (Scale 1:24,000, 1951, photorevised 1980), the site is located at an elevation of about 2840 feet above mean sea level. Site specific topography is relatively flat. Local area topography is sloped toward the southeast. Topography in the general vicinity is about 2900 feet at its highest point west of the city.

Geology/Soils

The subject site lies in an area consisting of moderately deep, nearly level to moderately steep, well-drained clayey soils underlain by shale. These soils formed in slightly acid to mildly alkaline material derived from clay shale. The soils are part of the Pierre Series (PrA). PrA soils are found on uplands with 0 to 2 percent slopes, mainly in the irrigated parts of the county. This soil has low fertility and poor tilth. It has a low or very low available water capacity (0.08-0.12). Permeability is very slow (< 0.06), and runoff is also slow. The PrA classification also has a high shrink-swell potential and low salinity. Most areas are in native grass and are used for rangeland. Wheat is the main dryfarmed crop in the few cultivated areas. Native vegetation is mainly mid and short grasses. Depth to bedrock is reported to lie between 20 and 40 feet beneath the surface.

When reviewing potential site hazards, use a methodical approach!

Hydrology/Floodplains

According to interviews conducted at the local Brentwood office of the Bureau of Land Management, the water table in most of the Brentwood Area is very low and should pose no constraint on surface development. Subsurface development (i.e. basements) are not suited for the area due to the high shrink-swell potential of the soils. Surface water runoff is slow. Potable water is supplied by the city through two municipal wells reported to be about 5,000 feet deep (interview with Robert Underwood, P.E., Bureau of Land Management and native of Brentwood, Colorado).

According to Mr. Robert Underwood, P.E., Brentwood office of the BLM, the city gets its water from wells and any spillage they may permeate the surface is of no consequence for two reasons: 1) the offsite wells extend some 500 feet below the surface; and 2) groundwater is not a source of potable water throughout the entire area.

15

The Site Investigation and Review of Hazards also presents information on...

- ☑ **Onsite and offsite findings required in each area reviewed**
- ☑ **Aboveground storage tanks (ASTs)**
- ☑ **Underground storage tanks (USTs)**
- ☑ **Chemicals**
- ☑ **PCBs**

The site was not identified as lying in a floodplain. Due to the low permeability of the soils and depth to groundwater it is highly unlikely that any surface spills that may have occurred at the subject site could pose negative impacts to groundwater.

Aboveground Storage Tanks (ASTs)

Several empty and near empty (operational and nonoperational) ASTs were identified onsite (see figure 2 in Maps Appendix).

The site's operational tanks include two-500 gallon tanks for unleaded fuels and one-600 gallon tank for diesel fuel. Several nonoperational ASTs (three to five) were noted on or immediately adjacent to the subject site. On the day of our investigation, no leaks were observed around any of the ASTs.

The ASTs located at the front of the structure (SW corner) had gauges on them (see site photos in the appendices). These were the ASTs connected to the fuel pump. Only one fuel pump was observed on the day of our investigation.

No spill containment devises were observed, and as previously mentioned the State does not appear to be concerned with ASTs in the rural farmland areas of Colorado.

Underground Storage Tanks (USTs)

According to our site reconnaissance, no evidence of USTs was noted on the grounds of the site, and no state database was located.

Chemicals

Record reviews and personal interviews indicate that the site's previous owners conducted the bulk storage and distribution of petroleum related products onsite. Several empty and near empty 55 gallon drums and miscellaneous containers were identified onsite (see Figure 2 in Maps Appendix). As most of the 55 gallon drums were unlabeled, their contents are unknown. Some of these drums and containers may have contained chemicals. No leaks or spills were noticed.

PCBs

It appears as though no transformers are located directly on the subject property, however two were identified on poles that line the rear and a portion of the side of the site. Due to their height above ground, labels could not be seen.

16

When ASTs and USTs are not found during a database search it does not mean they might not exist! It means that if they exist they were not reported to state agencies.

The Site Investigation and Review of Hazards also presents onsite and offsite findings for each area reviewed...

- ☑ **PCBs (contd.)**
- ☑ **Lead-based Paint**
- ☑ **Hazardous Substances**
- ☑ **Landfills**
- ☑ **Data references and interview results are crucial, and cannot be overemphasized if the potential hazard may impact the subject site**

As mentioned previously, there are no regulatory offices in Brentwood; hence there was no readily available source for further investigation.

The pole-mounted transformers, located near the property, did not show any evidence of staining on or surrounding the units during our site investigation. Transformers are a potential environmental concern because some types contain a hazardous substance in the dielectric fluid called Polychlorinated Biphenyls, or PCBs, which can leak from the transformer unit and impact the surrounding environment. Based upon a law promulgated by the EPA in 1985, the deadline for replacing or modifying transformers was October 01, 1990. Thus, the PCB content of these transformers is assumed to be no greater than 500 parts per million.

Lead-Based Paint

Exterior painted surfaces are relatively new. Based on an examination through a window of the office area, interior walls consist of wood paneling. CEI was unable to visually inspect the storage/workshop area, but based on the type of structure (prefab metal) believes it is unlikely that lead-based paint poses a significant impact to the site. CEI does not recommend that a lead-based paint survey be performed at this property. No exterior signs of blistering or peeling were observed.

Hazardous Substances

Based upon the site visit; interviews with local and state officials; and review of federal, state, and local files, evidence of current or prior release of petroleum or hazardous materials/wastes was not found on the property. No RCRA generators were identified within one mile of the site. As previously mentioned, several empty and near empty ASTs, bulk storage containers, and 55-gallon drums were identified onsite (see Figure 2 in Maps Appendix). As most of the 55-gallon drums were unlabeled, their contents are unknown. Some of these drums and containers may have contained hazardous substances.

Environmental litigations or administrative actions related to released or threatened releases of hazardous substances or petroleum products are not known to exist in regard to the site or abutting properties.

Landfills

No solid waste landfills were reported on, adjacent, or within one mile of the subject property.

17

If drums or containers are observed that may contain hazardous substances, the recommendation should be to sample their contents prior to any removal, just in case special disposal procedures must be utilized!

The Site Investigation and Review of Hazards also presents onsite and offsite findings for each area reviewed...

- ☑ Pits, sumps, drywells and catchbasins
- ☑ Stormwater drainage
- ☑ Drinking water
- ☑ Asbestos-containing materials
- ☑ Radon
- ☑ Asbestos and lead-based paint issues are typically not required in a basic Phase I ESA
- ☑ Findings from each area reviewed constitute the summary and recommendations section of the ESA report

Pits, Sumps, Drywells, and Catchbasins

Based upon observations and data collected during this survey, the site does not contain private wells, a septic system, sumps, or groundwater monitoring wells.

Stormwater Drainage

Stormwater drainage across the site appears to travel from north to south.

Drinking Water

Drinking water is supplied through two municipal wells. Water department officials indicated there were no known lead problems associated with these wells.

Asbestos-Containing Materials

An asbestos inspection of the interior and exterior areas of the JDE Oil Site was scheduled, and conducted on September 27, 1995. Six samples were taken and submitted for laboratory analyses. All samples were found not to contain asbestos (see complete Asbestos Survey and photos in the appendices)

In most states, conducting asbestos, lead-based paint, and radon surveys requires special state certifications.

Radon

CEI did not identify any concerns in this area. And as this is a small commercial type site and soils in the area are not conducive to subsurface development, the structure has a slab-on-grade concrete foundation, further limiting the significance of negative impacts potentially posed by radon. CEI does not recommend that a radon survey be performed at this property.

18

The Special Resource issue applies to sites that fall within specific agency age and size rules (typically 50 years, 50 acre undeveloped and/or 1 acre developed) and includes...

- ☑ **Historic resources**
- ☑ **Archaeological resources**
- ☑ **Endangered species**
- ☑ **Recreational areas/farmlands**
- ☑ **Sole-source aquifers**
- ☑ **Wetlands**
- ☑ **Resources that are obviously not applicable, such as Coastal and Wilderness Resources in this ESA, need not be mentioned; when it is not so obvious, a brief statement is required.**

7.0 Review of Special Resources

Based upon discussions with The Nature Conservancy (TNC) and the Advisory Council on Historic Preservation (ACHP), the property does not fall within their review guidelines, and on the day of our investigation no special resource issues were identified on, or adjacent to, the subject site.

Historic Resources (Historic Preservation Act)

The site consists of a prefab metal building of somewhat recent construction and has no historic significance.

Archaeological Resources

There are no known archaeological resources on or adjacent to the site.

Endangered Species

There are no aquatic refuges, shrubby grasslands, herbaceous woodlands, woody wetlands, woodlands/forests or other type of habitat areas on, or adjacent to, the site.

Farmlands (Farmlands Protection Act)

Historical records and aerial photos show possible evidence of prior site usage with respect to agricultural operations.

Recreational Areas

No major public recreation areas are on, or adjacent to, the site.

Land Use (Federal Land Policy and Management Act)

Land uses in the area are a mix of agricultural, residential, and small commercial establishments. The site is compatible with its surrounding service area.

Sole-Source Aquifers (Safe Drinking Water Act)

Federally designated sole-source aquifers were not identified within the site area's groundwater basin.

Wetlands/Wild and Scenic Rivers

No protected wetland areas are on, or immediately adjacent to, the site. There are no designated wild and scenic rivers in close proximity to the site.

19

In densely populated areas, special resource issues, other than historic sites, are typically not applicable!

The Summary and Recommendations portion includes...

- ☑ Summary of site history, surrounding area, and site-specific findings
- ☑ Hazard and special resource summary
- ☑ Recommendations for each hazard (if any)
- ☑ Recommendations for each special resource (if any)
- ☑ Recommendations for Phase II sampling (if needed)
- ☑ An estimate of environmental risk (low, moderate, high)
- ☑ Additional concerns (non-environmental hazards, e.g. structural damage)

8.0 Summary and Recommendations

Summary

Based upon site survey observations, interviews, and historic and present usage of the property, it appears that the site has been utilized primarily as an office/commercial property since its construction in 1980.

Record reviews and personal interviews indicated that the site is used to store and distribute petroleum-related products to farmers in the area.

A Comprehensive Asbestos Survey was conducted, and no asbestos-containing materials were identified onsite.

The surrounding area is primarily agricultural, low-density residential, with some minor commercial establishments along Arapahoe Avenue.

No industrial facilities were identified in the immediate site area.

Several empty and near empty ASTs, bulk storage containers, and 55-gallon drums were identified onsite (see Figure 2 in Maps Appendix). Most of the 55-gallon drums were unlabeled and their contents are unknown.

This section is a synopsis of all issues that may pose negative impacts on the subject site!

Recommendations

1) CEI recommends the empty 55-gallon drums, bulk storage containers, non-operational ASTs, and miscellaneous debris be removed from the site and disposed of properly.

2) As some of the drums, containers, and ASTs identified may have contained hazardous substances, CEI recommends that any item still containing fluids be sampled prior to disposal to determine whether it must be disposed of in accordance with hazardous materials regulations.

Based upon data collected during the site visit, interviews, and file reviews, the *environmental risk for the subject site is expected to be moderate to high.* There were no other concerns identified during the site investigation.

20

The Warrants are very important!

- ☑ They protect the inspector by limiting the ESA to a description of readily available environmental information obtained on specific dates and times
- ☑ They limit an inspector's legal and title search liability
- ☑ They acknowledge changes in site status due to the passage of time
- ☑ They exclude subsurface hazards and related issues
- ☑ Failure to include warrants may lead to problems if: illegal dumping occurs after an inspection; autos onsite cover spills or wells; inaccessible buildings or areas contain hazardous materials; database searches list the site after a search was completed, etc.

9.0 Warrants

This warranty is in lieu of all other warranties either express or implied. Conclusions set forth in this report are based upon, and limited by, the government data and other information available to CEI. While CEI has used reasonable care to avoid reliance upon faulty or incomplete information, CEI is not able to verify the accuracy of all data and information provided by governmental entities and third parties. Therefore, CEI is not responsible for any conclusion contained in this report that is based, in whole or in part, upon inaccurate data obtained from third parties.

It should be noted that all surface environmental assessments are inherently limited in the sense that conclusions are drawn and recommendations developed from information obtained from limited research and site evaluation at a specific time. The passage of time may result in a change in environmental circumstances at this site and surrounding properties, or hazardous materials beneath the surface or covered by debris may be present but undetectable during a site investigation. Verification of subsurface conditions, including the hazard potential of buried or covered debris, is beyond the scope of this investigation.

All asbestos and lead-based paint surveys are inherently limited in the sense that conclusions are drawn and recommendations developed from information obtained from limited site sampling at a specific time. The passage of time may result in deterioration and/or damage to non-friable Asbestos Containing Materials, or a lead-based paint may be used in areas where these materials did not exist and/or areas where no sampling was conducted.

The information herein is for the exclusive use of the John Doe Enterprises, Inc. and CEI. CEI is not responsible for use of this information by any other parties. This company is not responsible for the independent conclusions, opinions, or recommendations made by others based on the field exploration, sampling, and laboratory analyses presented in this report.

CEI does not provide professional legal or title insurance services and makes no guarantee, explicit or implied, that the listing which was reviewed represented a comprehensive delineation of past site ownership or tenancy. The work performed in conjunction with this assessment and the data developed are intended as a description of available information at the dates and locations given.

21

As it pertains to an ESA, a warrant is defined as: a promise that a certain statement of facts, in relation to a subject site, is as it is represented to be!

The ESA Appendices/Exhibits...

☑ Note that site maps, site photos, aerial photos and environmental database searches are site-specific and confidential in nature.

☑ At a minimum, the following information should always be included:

☑ Site maps

☑ Site photos

☑ Aerial photos

☑ Database summary

☑ Data references

Sample Phase I ESA

Appendices/Exhibits

- Site Maps
- Site Photos
- Aerial Photos
- Regulatory Database Summary
- Data References

22

Section dividers (title flysheets) are typical in ESA appendices...

- ☑ **The inspector should announce each exhibit and/or appendices section with a flysheet.**

Site Maps

The Site Map...

- ☑ **The title block and frame should be a boilerplate with blanks for project-specific information.**

- ☑ **Notice that this map is a copy of a USGS Quad. This scale orients the reader to the location of the site relative to the general region.**

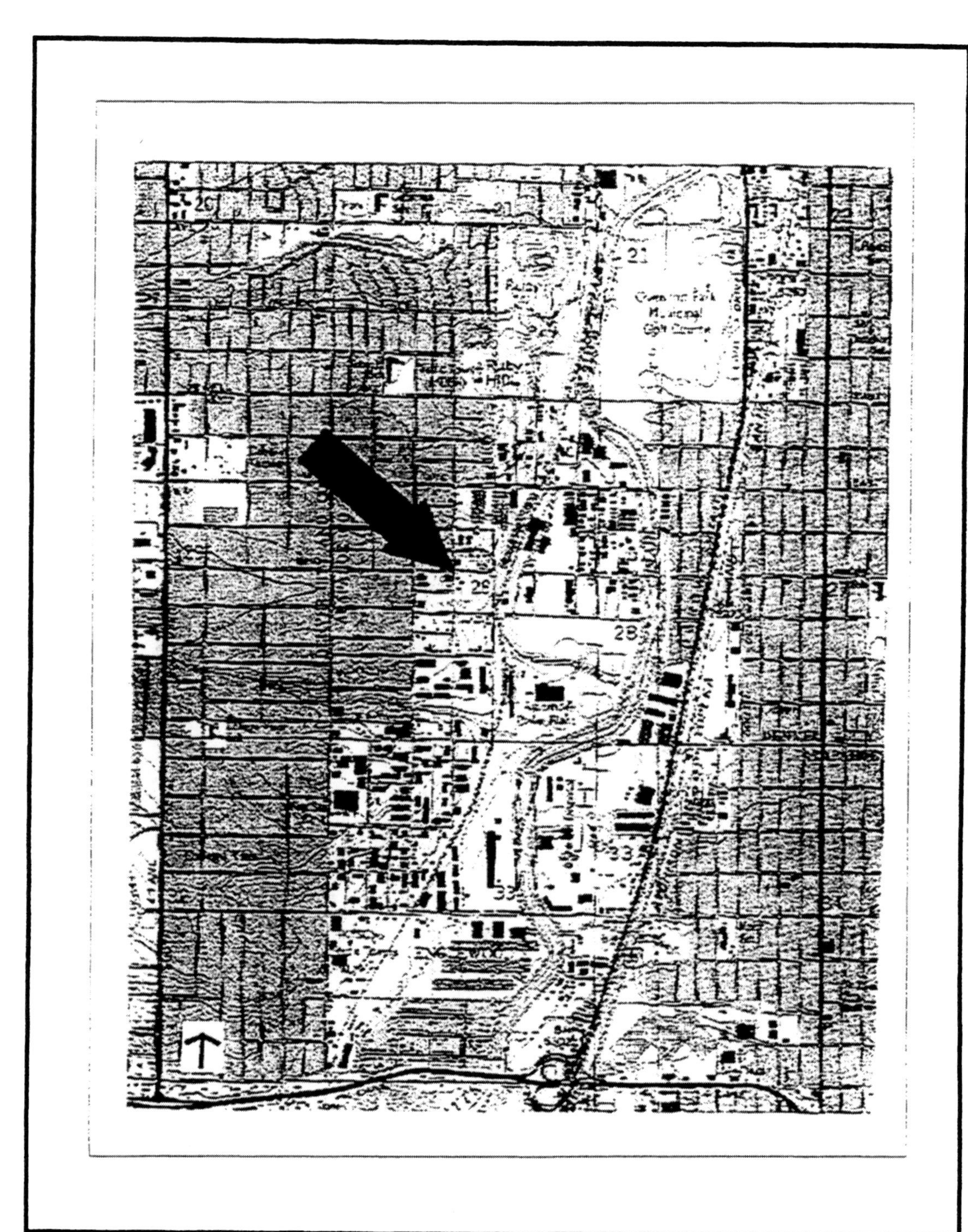

PREPARED BY:

PROJECT DATE:

EXHIBIT NUMBER:

USGS 7.5 Minute
Topographic Quadrangle
Aurora South

Coopers Environmental

SCALE: 1:6400

The Site Plan...

- ☑ As with the site map, the title block and border are boilerplate for an 8.5" x 11" presentation.
- ☑ The site plan may have to be self-generated using a CAD or graphics program.
- ☑ Note the location of potential hazards and important site characteristics on the plan.

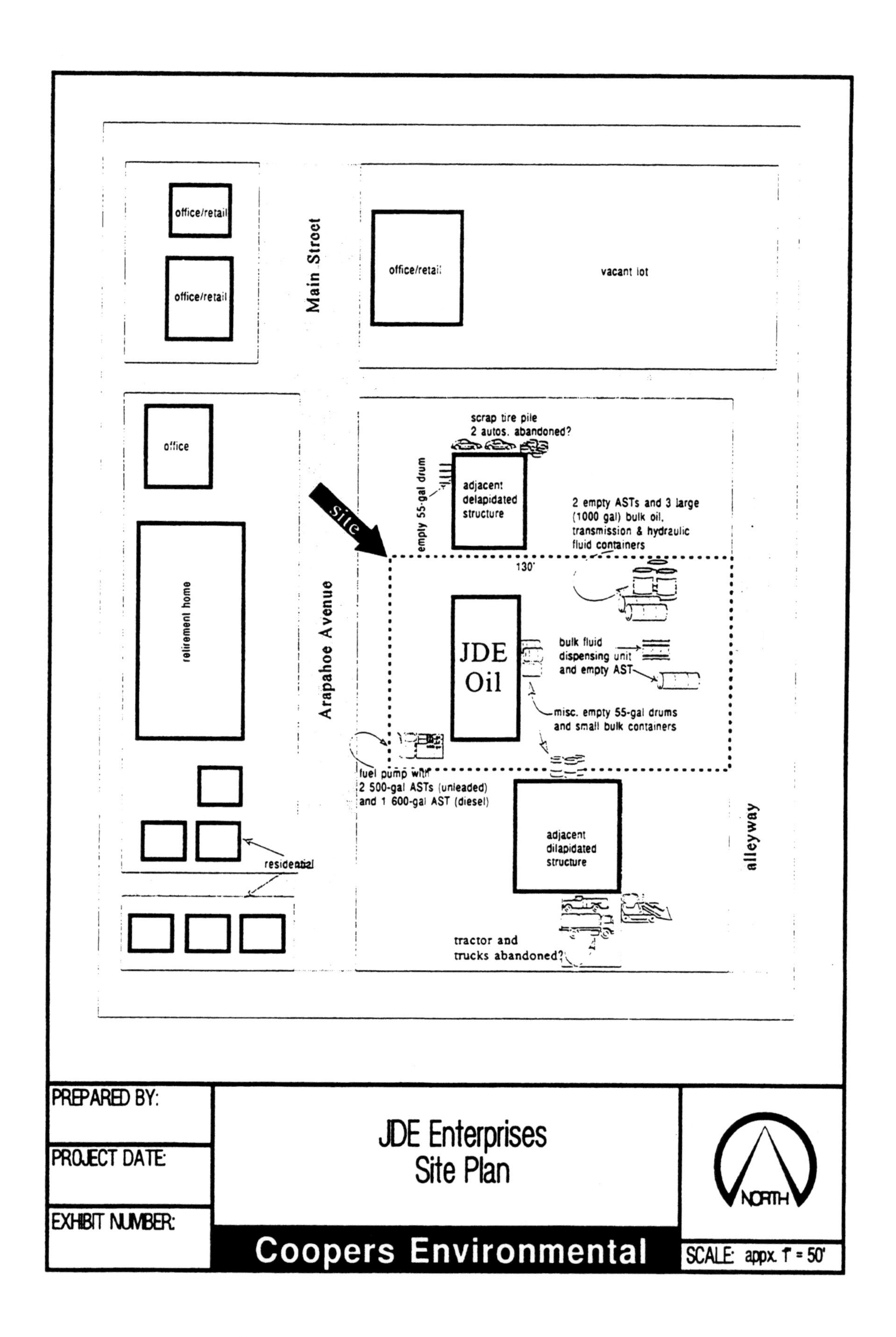
office/retail
office/retail
Main Street
office/retail
vacant lot
office
scrap tire pile
2 autos. abandoned?
empty 55-gal drum
adjacent
delapidated
structure
site
2 empty ASTs and 3 large
(1000 gal) bulk oil,
transmission & hydraulic
fluid containers
130'
retirement home
Arapahoe Avenue
JDE
Oil
bulk fluid
dispensing unit
and empty AST
misc. empty 55-gal drums
and small bulk containers
fuel pump with
2 500-gal ASTs (unleaded)
and 1 600-gal AST (diesel)
adjacent
dilapidated
structure
alleyway
residential
tractor and
trucks abandoned?
PREPARED BY:
PROJECT DATE:
EXHIBIT NUMBER:
JDE Enterprises
Site Plan
NORTH
Coopers Environmental
SCALE: appx. 1" = 50'

Section dividers (title flysheets) are typical in ESA appendices...

- ☑ **The inspector should announce each exhibit and/or appendices section with a flysheet.**

Site Photos

The Site Photos...

- ☑ The presentation of photos should also be done with boilerplate title blocks and borders.
- ☑ Caption each photo to orient the reader to the site.
- ☑ Exterior photos are typically presented first.

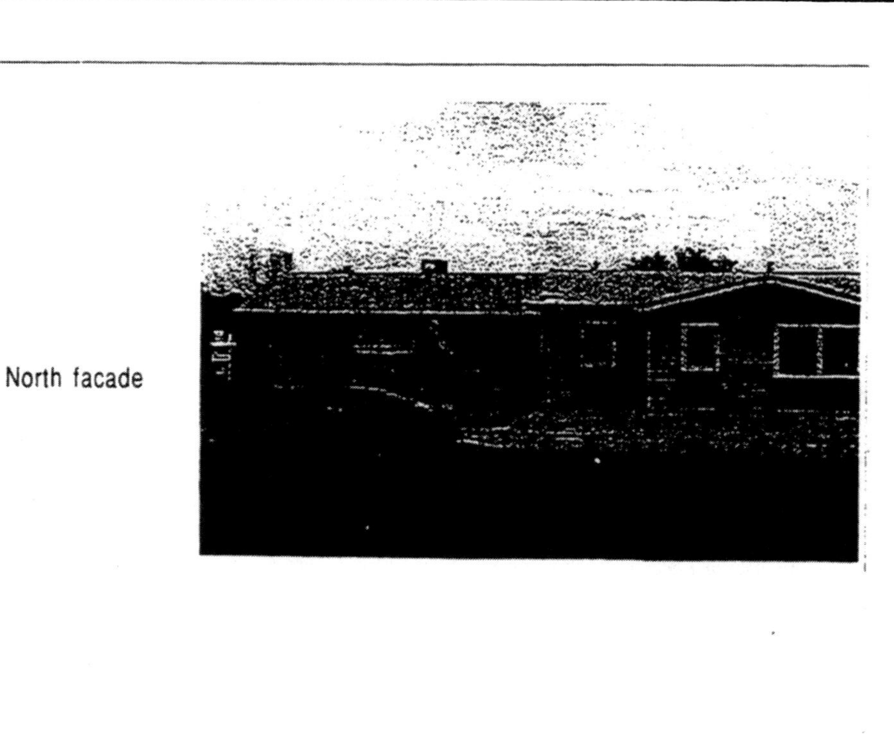

North facade

South side of site looking north

PREPARED BY:	JDE Enterprises Site Photos	
PROJECT DATE:		
EXHIBIT NUMBER:	**Coopers Environmental**	SCALE: n/a

The Site Photos...

☑ **The inspector should ensure enough photos are taken to give a reader unfamiliar with the site a complete picture of the site.**

North side of site
facing south

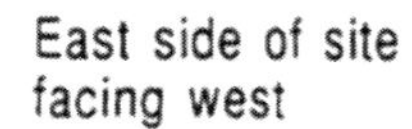

East side of site
facing west

PREPARED BY:	JDE Enterprises Site Photos	
PROJECT DATE:		
EXHIBIT NUMBER:	**Coopers Environmental**	SCALE: n/a

The Site Photos...

☑ **The inspector should take interior site photos with the intent of showing material surfaces, conditions of building materials, and storage/disposal activities (if any).**

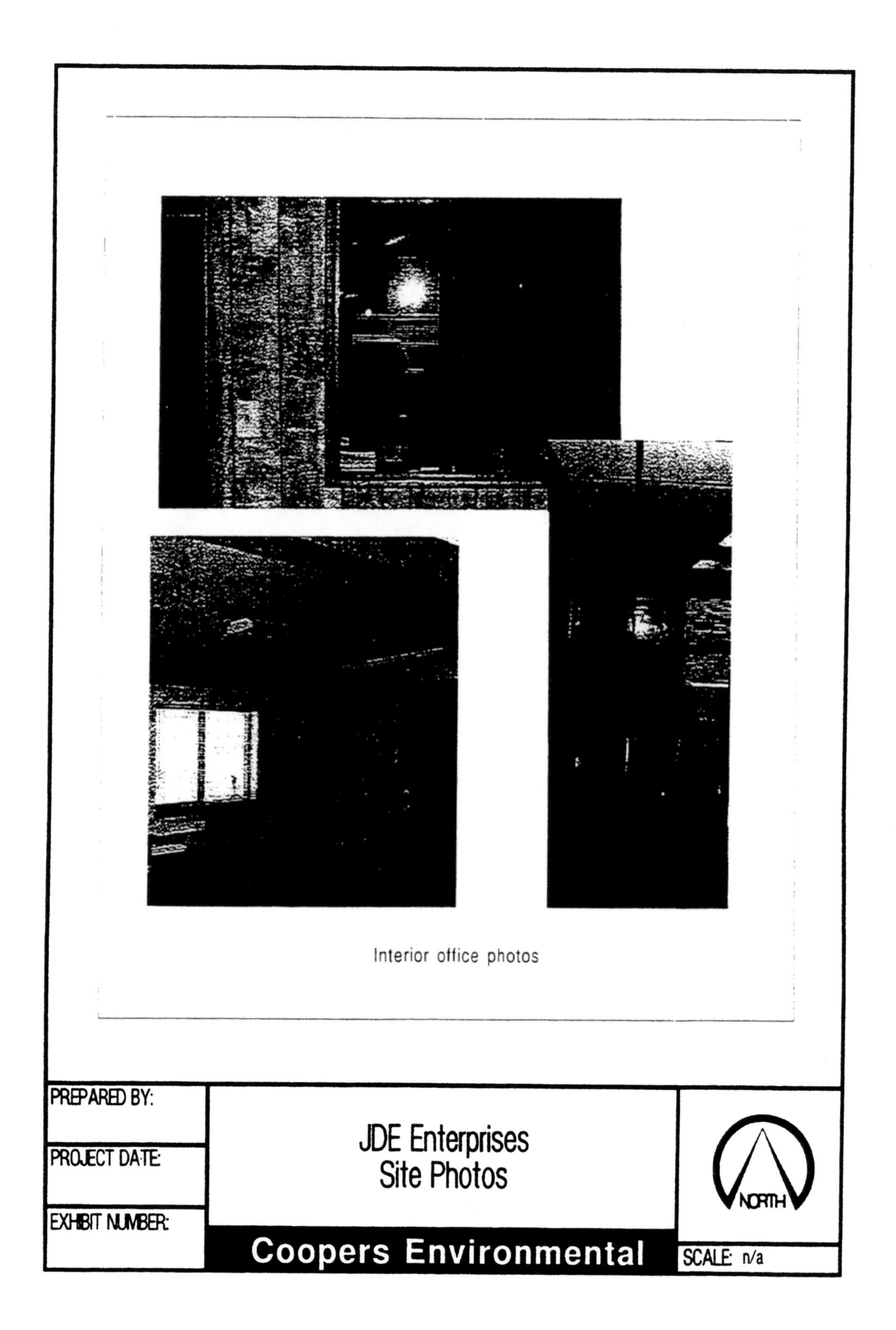

Interior office photos

Section dividers (title flysheets) are typical in ESA appendices...

- ☑ **The inspector should announce each exhibit and/or appendices section with a flysheet.**

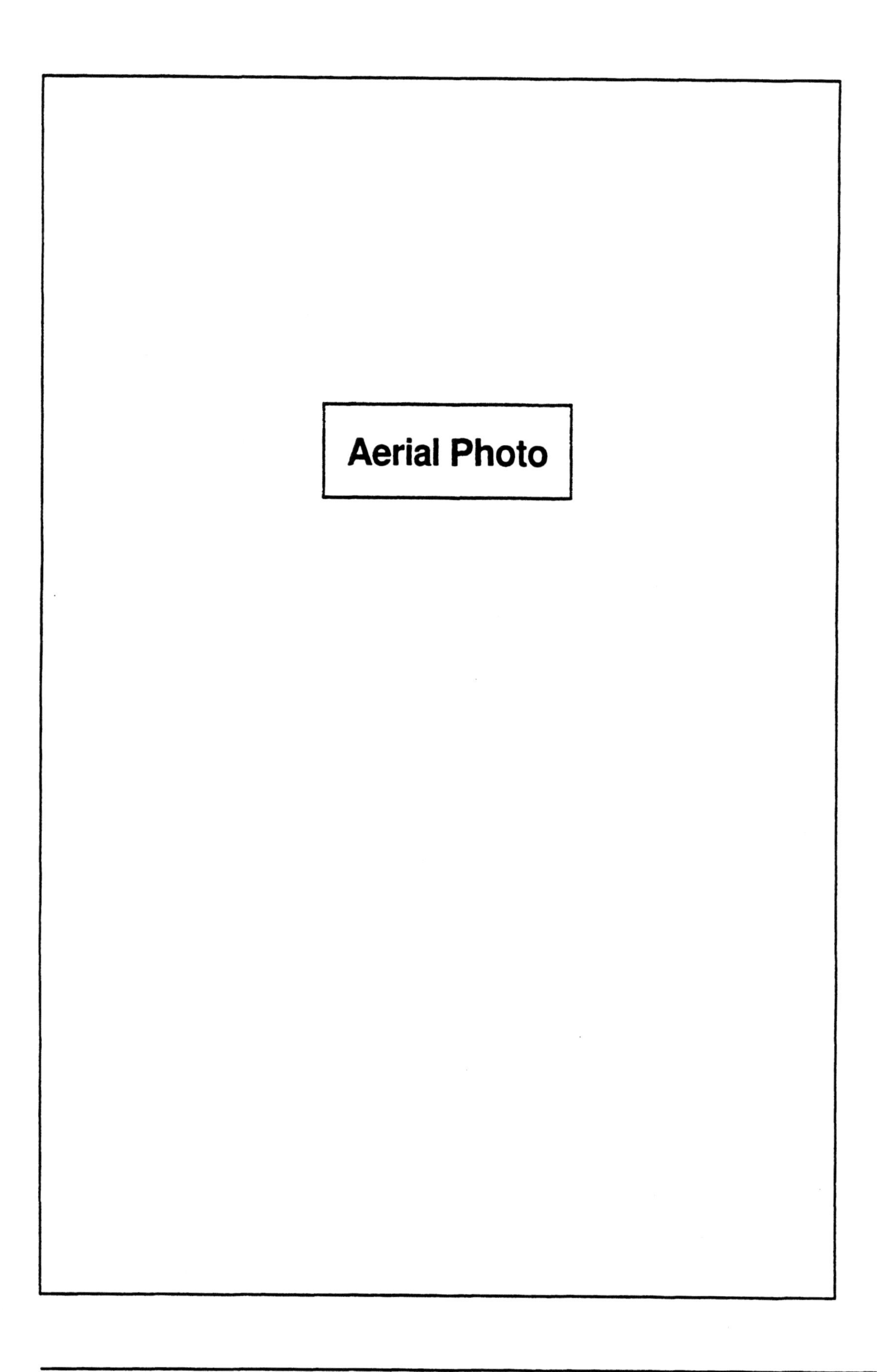
Aerial Photo

The Aerial Photos...

- ☑ The presentation of aerial photos should also be done with boilerplate title blocks and borders.
- ☑ When possible, the inspector should present three aerial photos (an historic aerial, a post-development aerial, and a current aerial).
- ☑ The actual site should be pointed out to the reader.

PREPARED BY:	JDE Enterprises 1962 Aerial Photo	NORTH
PROJECT DATE:		
EXHIBIT NUMBER:	**Coopers Environmental**	SCALE: appx. 1" = 200'

Section dividers (title flysheets) are typical in ESA appendices...

- ☑ **As with the other sections, the inspector should announce each exhibit and/or appendices section with a flysheet.**

Regulatory Database Summary

The Environmental Database Summary...

- ☑ If the inspector conducts his/her own regulatory database searches, he/she should develop a boilerplate format that provides a quick summary of pertinent information.

- ☑ Try to restrict the summary sheet to one page when possible.

Phase I ESA Database Summary

PROPERTY INFORMATION	CLIENT INFORMATION
Project Name: JDE OIL SITE 124 E. Arapahoe Avenue Brentwood, Colorado	JDE Enterprises 4000 Main Street, Suite 300 Denver, Colorado 80112

ENVIRONMENTAL RISK DISTRIBUTION SUMMARY Agency/Database - Type of Records			within 1/8 mile	1/8 to 1/4 mile	1/4 to 1/2 mile	1/2 to 1 mile
A) RISK SITES searched to 1 mile:						
US EPA	NPL	Sites designated for Superfund Cleanup by EPA	0	0	0	0
US EPA	RCRIS	Facilities that Treat, Store and/or Dispose of Haz Waste	0	0	0	0
STATE	WQARF	Sites prioritized by state for cleanup	0	0	0	0
B) RISK SITES searched to 1/2 mile:						
US EPA	CERCLIS	Sites under Review by US EPA	0	0	0	-
STATE	LUST	Sites with Leaking Underground Storage Tanks	0	0	0	-
STATE	SWLF	Sites permitted as Solid Waste Landfills, Incinerators, or Transfer Stations	0	0	0	-
C) RISK SITES searched to 1/4 mile:						
STATE	UST	Sites with Registered Underground Storage Tanks	0	0	-	-
D) RISK SITES searched to 1/8 mile:						
US EPA	ERNS	Sites with Previous Haz. Materials Spills	0	-	-	-
US EPA	RCRIS	Sites that generate Large Quantities of Haz Waste	0	-	-	-
US EPA	RCRIS	Sites that generate Small Quantities of Haz Waste	0	-	-	-

This geographic database search meets the American Society of Testing Materials (ASTM) standards for a government records review. A (-) indicates search distance exceeding ASTM search parameters.

LIMITATION OF LIABILITY

Client proceeds at its own risk in choosing to rely on this database search, in whole or in part, prior to proceeding with any transaction. LTRE cannot be an insurer of the accuracy of the information, errors occurring in conversion of data, or for client's use of data.

Sample Only

Section dividers (title flysheets) are typical in ESA appendices...

- ☑ **Announce each exhibit and/or appendices section with a flysheet.**

Data References

The Data References...

☑ **The inspector should organize information such that the sources of exhibits, studies, and interviews are listed.**

Data References

Quadrangle Map

United States Geological Survey, 1962, 7.5 Minute Series Topographic Map, Brentwood South Quadrangle, Photo revised 1980.

Plans, Studies, Maps, and Reports Reviewed

Arapahoe County, Colorado Soil Survey

Arapahoe County, Colorado Soil Map

Aerial Photographs

Bureau of Land Management, Brentwood, CO, mid-1980s.

Assessors Data and Maps

Arapahoe County Recorder and Tax Assessors Office, Denver, CO

Agencies Interviewed

City of Brentwood Water Department

Bureau of Land Management, Brentwood, CO

Arapahoe County Planning, Denver, CO

Arapahoe County Tax Assessor

Appendix B

Phase I ESA Inspection Checklist

Due Date:______________ Job No.______________

PHASE I ESA INSPECTION CHECKLIST

Project Manager ____________
Date of inspection ____________

Instructions: *This ESA checklist should be completed during site reconnaissance activities for all properties, except single-family residential, based on readily available information. Following completion, the inspector should use this data along with information obtained from the environmental database and related site information, to compile the ESA. The data should be used to determine any appropriate and applicable follow-up activity at the property. The checklist should also be included in the appendices of the ESA.*

■ SITE/CONTACT DATA

Project Address:

City: County:

State: Zip:

Site Mgr./Contact Name:

Phone: FAX:

Mgt. Office Address:

City: State: Zip:

■ PROJECT MANAGER/CONSULTANT DATA

Firm Name:

Environmental Inspector:

Phone: FAX:

Address: City: State: Zip:

Are Confidentiality Agreements on file? Yes ☐ No ☐

■ SITE DESCRIPTION

Site Category
- ☐ Residential
- ☐ Commercial
- ☐ Vacant or Undeveloped Land

Current Property Use
- ☐ Single-Family
- ☐ Multi-Family
- ☐ Retail
- ☐ Office
- ☐ Industrial, *specify type:*______________
- ☐ Agricultural, *specify type:*___ _________

■ PROPERTY SPECIFICATIONS

Total Square Feet	# Units	Total Acreage
Legal Description:		
Current Property Status	Vacant ☐	Occupied ☐
General Description:		
Name the two nearest crossroads:		
Give your best estimate of the age of the onsite buildings:		
Does site have over 50 acres of generally undeveloped land?	Yes ☐	No ☐
Does site have over 1 acre of naturally vegetated land, excluding improvements?	Yes ☐	No ☐
Provide a general site description:		
Adjacent Properties Site Description:		
North:		
South:		
East:		
West:		

■ SITE HISTORY & REFERENCES *(circle all that apply)*

Past Use of Property:		Site Category:
Single-Family		Residential
Multi-Family		Commercial
Retail/Office		Vacant/Undeveloped Land
Industrial, *specify type*:	Agricultural, *specify type*:	
Site/Local government interviews conducted?	Yes ☐	No ☐
50-Year Chain-of-Title completed/submitted?	Yes ☐	No ☐
Thıee Historic Aerial Photographs obtained?	Yes ☐	No ☐
7.5 Minute USGS Quad Map obtained/submitted?	Yes ☐	No ☐
Site and vicinity plans obtained/submitted?	Yes ☐	No ☐
Comments:		

■ SITE RECONNAISSANCE

Were any of the following environmental issues observed on or immediately adjacent to the property while conducting onsite reconnaissance activities? Yes ☐ No ☐

Environmental Issue	Yes/No	Notes
Industrial plants	Yes☐ No☐	
Waste transporters	Yes☐ No☐	
TSD facilities	Yes☐ No☐	
Vegetative stress	Yes☐ No☐	
Storage tanks	Yes☐ No☐	
Stained soil	Yes☐ No☐	
Historic sites	Yes☐ No☐	
Wetlands/floodplains	Yes☐ No☐	
Wild/scenic rivers	Yes☐ No☐	
Dump areas	Yes☐ No☐	
Lead paint	Yes☐ No☐	
Lead piping	Yes☐ No☐	
Asbestos materials	Yes☐ No☐	
Water wells	Yes☐ No☐	
Unique wildlife species	Yes☐ No☐	
Discarded batteries	Yes☐ No☐	
Fluorescent fixtures	Yes☐ No☐	
Oil/Gas drums	Yes☐ No☐	
Coastal areas	Yes☐ No☐	

Other Findings/Comments:

■ REGULATORY DATABASE SEARCHES

Were any of the following environmental database searches conducted? Yes ☐ No ☐

Environmental Databases	Sites within 1 mile	Sites within .25 miles
CERCLIS (Superfund)	Yes ☐ No ☐	Yes ☐ No ☐
ERNS	Yes ☐ No ☐	Yes ☐ No ☐
RCRIS	Yes ☐ No ☐	Yes ☐ No ☐
RCRA	Yes ☐ No ☐	Yes ☐ No ☐
NPL	Yes ☐ No ☐	Yes ☐ No ☐
FINDS	Yes ☐ No ☐	Yes ☐ No ☐
TSD	Yes ☐ No ☐	Yes ☐ No ☐
UST	Yes ☐ No ☐	Yes ☐ No ☐
LUST	Yes ☐ No ☐	Yes ☐ No ☐
Solid Waste Landfills (open ☐ abandoned ☐)	Yes ☐ No ☐	Yes ☐ No ☐

Others, please specify:

■ ENVIRONMENTAL HAZARDS INVESTIGATIONS

USTs/ASTs

Was there any evidence of the presence of current or former service stations, or of commercial or industrial activities that suggest that underground storage tanks may be located on the property? Yes ☐ No ☐

Is there any evidence that ASTs or USTs may be located on the property? Yes ☐ No ☐

Landfills/Dumping Activities

Is there any evidence that a landfill, dump, wastepile, wastewater lagoon, or other land disposal activity is currently present on this property? Yes ☐ No ☐

Hazardous Substances

Is there any evidence that substances such as paints, solvents, acids, gases, flammables, poisons, or other chemicals are currently used on the property? Yes ☐ No ☐

Is there any evidence that radiological materials such as found in laboratories, medical equipment, or industrial operations may be used on the property? Yes ☐ No ☐

Is there any evidence that past operations used hazardous substances that may have been released into the environment, or that the property may have been used for dumping, landfilling, or disposing of hazardous materials in the past? Yes ☐ No ☐

■ HAZARDS INVESTIGATIONS (contd.)

Pesticides

Is there any evidence that pesticides (including insecticides, fungicides, and rodenticides) have been manufactured or used on the property? Yes ☐ No ☐

Pits, Sumps, Wells

Does the site use an onsite well? Yes ☐ No ☐

Does the site use an onsite septic system? Yes ☐ No ☐

Are there any onsite monitoring wells? Yes ☐ No ☐

Asbestos-Containing Materials Is this applicable to SOW? Yes ☐ No ☐

Was there any evidence of insulation or fire retardant materials such as pipe wrap and ceiling spray within the buildings on the property? Yes ☐ No ☐

Are there any residential structures onsite that were built prior to 1978? Yes ☐ No ☐

Were any Asbestos-Containing Materials (ACMs) observed on the property? Yes ☐ No ☐

Were sprayed-on insulating materials observed on piping/fireproofing areas? Yes ☐ No ☐

Were photos taken of suspected ACMs? Yes ☐ No ☐

Has an asbestos survey been conducted? Yes ☐ No ☐

Lead Paint Is this applicable to SOW? Yes ☐ No ☐

Was any visible evidence of paint peeling, cracking, or flaking noticed? Yes ☐ No ☐

Was there any evidence of lead paint containers being stored onsite? Yes ☐ No ☐

Was interview conducted with site maintenance personnel? Yes ☐ No ☐

Was there any other visible evidence of lead paint on any site structures? Yes ☐ No ☐

Polychlorinated Biphenyls (PCBs)

Were any transformers, electrical devices, fluorescent ballasts, or equipment stored and/or observed on the property labeled as containing PCBs? Yes ☐ No ☐

Was there any evidence of PCB contamination or leaks observed? Yes ☐ No ☐

Radon Is this applicable to SOW? Yes ☐ No ☐

Are you aware of elevated levels of radon present in the site vicinity? Yes ☐ No ☐

Is there reason to suspect radon to be a problem at the site? Yes ☐ No ☐

Has radon screening been conducted at the site? Yes ☐ No ☐

Does the local health department have any information on radon onsite? Yes ☐ No ☐

■ ADJACENT SITES

Do any of the sites adjacent to the subject site exhibit any of the potential issues identified in the hazards checklist above. If YES, explain:

■ HAZARDS INVESTIGATIONS (contd.)

Comments and explanation of YES Answers to Hazards Investigation Section:

■ SPECIAL RESOURCES

Archaeological Resources

Is there any reason to believe that there are or may be any archaeological resources on the property that have not been destroyed by human activity? Yes ☐ No ☐

Has this property recently been developed, excavated, mined, graded, or otherwise disturbed by human activity over all or nearly all of this area? Yes ☐ No ☐

Is this property covered or nearly covered by buildings and/or pavement (such as parking lots) that are less than 50 years old? Yes ☐ No ☐

Is this property located in a region known to contain significant fossil collections or remnants of organisms of a past geologic age? Yes ☐ No ☐

Coastal Resources

Are there any coastal areas, rivers, streams, springs, lakes, ponds, swamps, marshes, or other bodies of water on or immediately adjacent to the property, or does the property contain any area or areas containing natural vegetation greater than 1 acre in size? Yes ☐ No ☐

Is the propertv located within approximately 1,000 feet of the shoreline of the Atlantic or Pacific Oceans, Gulf of Mexico, other major saltwater tributary (e.g., Chesapeake Bay or San Francisco Bay) or Great Lakes? Yes ☐ No ☐

Is the area located within 1,000 feet of a shoreline and developed with industrial, commercial, or high-density residential uses which have eliminated or significantly degraded any existing coastal or beach resources? Yes ☐ No ☐

Historic Resources/National Landmarks

Is there any evidence that the property has a designated natural landmark or has buildings that are more than 50 years old or buildings that are less than 50 years old but may be considered historically significant? Yes ☐ No ☐

Are there any buildings or structures on the property? (A structure includes objects such as bridges or dams, which are created by humans and are permanently located on the property.) Yes ☐ No ☐

Are there any buildings or structures less than 50 years old that may be considered exceptionally significant in American history, architecture, archaeology, engineering, or culture (such as the work of a renowned architect or a structure associated with a very significant individual)? Yes ☐ No ☐

■ SPECIAL RESOURCES INVESTIGATIONS (contd.)

Historic Resources (contd.)

Could this property be considered a place where a significant event or pattern of events occurred such as a battlefield or prehistoric settlement? Yes ☐ No ☐

Could this property be considered part of a Historic District (over 50 years old)? Yes ☐ No ☐

If the District itself or the historical context that makes this District important is less than 50 years old, may it be considered exceptionally significant in American history, architecture, archaeology, engineering, or culture? Yes ☐ No ☐

Does this property contain a designated natural landmark (look for plagues, markers, or other indications of landmark status)? Yes ☐ No ☐

Has this property recently been developed, excavated, graded, or experienced significant disturbance by human activity over all or nearly all of this area? Yes ☐ No ☐

Recreational Areas

Does the property contain natural features which could be used for public recreational purposes? Yes ☐ No ☐

Sole-Source Aquifers

Is there any evidence that the property is located in an area that uses groundwater as the primary or sole source of drinking water? Yes ☐ No ☐

Wetlands/Endangered Species

Has any study been done on the property to assess wetlands, the presence of threatened and endangered species or critical and unique habitat? Yes ☐ No ☐

Does the property contain any area or areas containing natural vegetation greater than 1 acre in size? Yes ☐ No ☐

Are there any rivers or streams on or immediately adjacent to the property which appear to be in pristine or near pristine condition? Yes ☐ No ☐

Are the banks of the river or stream predominantly developed with industrial, commercial, or mid-to-high-density residential uses? Yes ☐ No ☐

Is there any reason to believe that the river or stream is protected under federal, state, or local law? Yes ☐ No ☐

Comments:

■ SUMMARY TABLE & RECOMMENDATIONS

Issue	Acceptability	Notes
Aboveground Storage Tanks (ASTs)	☐ acceptable ☐ unacceptable	
Underground Storage Tanks (USTs)	☐ acceptable ☐ unacceptable	
Database Findings	☐ acceptable ☐ unacceptable	
Special Resources	☐ acceptable ☐ unacceptable	
Solid Waste/Landfills	☐ acceptable ☐ unacceptable	
Polychlorinated Biphenyls (PCBs)	☐ acceptable ☐ unacceptable	
Asbestos	☐ acceptable ☐ unacceptable	
Lead Paint	☐ acceptable ☐ unacceptable	
Radon	☐ acceptable ☐ unacceptable	

■ ENVIRONMENTAL FINDINGS

☐ A. The subject site is not recommended for further study.

There is no evidence that the site is being impacted by any environmental hazards, and there is no evidence that the site is being impacted by any special resource issues.

☐ B. The subject site is recommended for further study in the following area(s):

☐ ASTs/USTs
☐ Hazardous Substances
☐ Landfills
☐ Asbestos
☐ PCBs
☐ Lead
☐ Radon

Archaeological Resources ☐
Coastal Resources ☐
Endangered Species ☐
Historic Resources/National Landmarks ☐
Recreational Areas ☐
Sole-Source Aquifers ☐
Wetlands ☐

Contact has ☐, has not ☐ been made with the appropriate federal, state, and local agencies to determine if the above area of concern is impacting the site.

☐ C. The subject site is recommended for further study. A study is ☐, or has been ☐ conducted to assess the following:______________________________

The study has ☐, has not ☐ been received by appropriate Federal and state agencies. The study has ☐, has not ☐ been approved.

☐ D. A Phase II ESA ☐ is, ☐ is not recommended, to assess:

■ ENVIRONMENTAL RISK

Based upon my findings, I consider this property to be a: ☐ no risk ☐ moderate risk ☐ low risk ☐ high risk

The risk rating above is based upon the following:

■ INTERVIEWS and CONTACT INFORMATION

Persons Interviewed & Date	Affiliation	Notes

Interview Comments:

■ OTHER ISSUES

☐ A. No other concerns were noted on or immediately adjacent to the subject site.

☐ B. Additional study should be conducted on

■ CERTIFICATION

I certify that to the best of my knowledge, the above information was obtained via customary and appropriate means of environmental inquiry and the data presented in this inspection are true and accurate based upon available information as of the inspection date; I personally conducted all site reconnaissance tasks of the subject property; and I have no undisclosed interest, present or prospective, in the subject property.

Inspector's Signature:_________________________ Date:__________

Appendix C

Asbestos Operations and Maintenance Plans

Asbestos Operations and Maintenance Plans

Operations and Maintenance (O&M) Plans are typically required when asbestos-containing materials (ACMs) are found onsite, and care must be taken to maintain them in a non-hazardous, non-friable state. O&M plans must be prepared by a certified AHERA-trained Asbestos Inspector/Management Planner in accordance with EPA protocol.

Operations and Maintenance (O&M) Plan

☐ If an O&M Plan is recommended, the site representative, based on the results of the Comprehensive Asbestos Survey (CAS), should determine whether an O&M Plan is to be prepared by the inspector.

The Six O&M Programs

There are six O&M programs that should be prepared in connection with a site's O&M Plan:

1. Notification Program

☐ An information program to inform employees and occupants at the site of where ACMs are located, and how and why to avoid disturbing the ACMs.

2. Monitoring Program

☐ An observation program to identify, assess, and document any changes in the condition of the ACMs.

3. Operations Program

- ☐ A permit system to control work practices and any other activities that might disturb the ACM, to avoid or minimize fiber release.

4. Recordkeeping Program

- ☐ A recordkeeping program to document all O&M activities.

5. Worker Protection Program

- ☐ A response program for medical and respiratory controls, if required.

6. Training Program

- ☐ An educational program designed to designate a responsible individual for implementing O&M activities and disseminating information to site personnel.

O&M Records

- ☐ O&M, abatement, and/or encapsulation plans and information are typically included in the appendices of the report and summarized in the Cover Letter, Hazards Investigations, and Summary and Recommendations sections of the ESA.

O&M Certification

- ☐ All inspectors and subinspectors must state in their ESA report that they are properly licensed and/or AHERA-certified to perform O&M Plans, and must provide evidence of such licenses and/or certificates.

Glossary of Terms

Glossary of Terms

A

Abandoned Well
A well whose use has been permanently discontinued due to lack of resource, need, or an inoperative physical condition.

Abatement
A decrease, reduction, diminution, or sometimes termination of a pollutant or the condition generating the pollution.

Aboveground Storage Tank (AST)
A tank device situated in such a way that the entire surface area of the tank is completely above the plane of the adjacent surrounding surface and the entire surface area of the tank can be visually inspected.

Abstract of Title
A summary or condensation of the essential parts of all recorded instruments which affect a particular piece of real estate; arranged in the order they were recorded.

Abut
To border on. When two adjacent properties share a common public easement, such as a highway, they are called "abutting" properties. Properties that share a common property line are adjoining or adjacent sites.

Access Agreement
An agreement allowing contractor/consultant access to a site throughout the course of a specific project period.

Accessory Structure
A structure on a site that is detached from the main facility and that has a use related, but incidental, to that of the main facility.

Accessory Use
In zoning, a use of land incidental to the major zoning classification for the property, such as parking lots in commercial zones and swimming pools in residential zones.

Act of God
An unanticipated grave natural disaster or physical phenomenon of an extraordinary, fated, and compelling character, the effects of which could not have been foreseen or avoided.

Activity Mapping
A method of recording and displaying performance data in relation to an environmental setting. The planner notes activities and their frequencies on site and on facility maps which, when completed, show operational behaviors; location of storage, treatment, shipping, disposal, and other activities; frequency of the storage, treatment, shipping, disposal and other activities at varying locations; waste movement; relation of people to facilities and projects; and people movement.

Adjacent Properties
The area immediately outside the property boundaries of the site. Specifically, adjacent means properties that would otherwise be contiguous but for an intervening ownership, strip, watercourse, gore, road or other feature.

Advisory Council on Historic Preservation (ACHP)
A federal agency established by Section 201 of the National Historic Preservation Act of 1966 (as amended). The Council reviews and comments on all federal actions affecting cultural resources under the authority of Section 106 of the National Historic Preservation Act. It also advises the President and the Congress on historic preservation matters; recommends measures to coordinate activities of federal, state, and local governments; and advises on the dissemination of information pertaining to such activities.

Air Pollution
The existence of contaminants in the air in concentrations too high to allow the normal dispersal ability of the air and that interfere directly, or indirectly, with human health, safety, welfare, or comfort, and/or inhibit the full use and enjoyment of property.

Amenities
In real estate, amenities refer to such circumstances (e.g., location, outlook, or access to a park, lake, highway, river) which enhance the location, aesthetic environs, or desirability of a particular site and which contribute to the enjoyment of the occupants.

Appraisal
An estimate, or opinion, of value. In the real estate business, formal appraisal reports are relied upon in important decisions made by mortgage lenders, investors, public utilities, governmental agencies, businesses, and individuals.

Aquifer
An underground rock formation composed of materials such as sand, soil, or gravel that can store and supply groundwater to wells and springs; a geologic formation, group of formations, or part of a formation. Most aquifers used in the United States are within about 1000 feet of the earth's surface.

Archaeological Site
Sites evidencing ancient living areas or ancient farming, hunting, gathering activities, or of burials and funerary remains, artifacts, and structures of all types. These sites usually date from prehistoric or aboriginal periods, or from historic periods of which only vestiges remain.

Asbestos
A noncombustible fibrous mineral used for fireproofing in industrial and construction applications. Prolonged exposure to airborne asbestos fibers through inhalation has been deemed to be extremely hazardous and can cause various forms of cancer and respiratory diseases.

Asbestos Standards
Regulations promulgated by the U.S. Occupational Safety and Health Administration (OSHA) in 1986 that require major reductions in the level of airborne asbestos fibers in work places, and that also prescribe a system of engineering controls and work practices related to asbestos. Actually two standards exist, one for General Industry, and one for the Construction Industry, with somewhat differing requirements.

Assessment
In environmental planning, the evaluation and appraisal of a project or property for the purpose of approximating and planning for the mitigation of adverse effects upon the surrounding environs. Also, used as a primary means of developing alternative approaches of completing a project, while simultaneously minimizing or avoiding negative impacts.

B

Baseline Data
A collection of information regarding the existing conditions of a specific and geographically defined environmental area, usually taking into account the built, natural, wildlife, human, cultural, and economic environmental characteristics.

Bedrock
A general term for the rock, usually solid, that underlies soil or other unconsolidated material.

Biota
Combined fauna and flora of any area or geological period.

Buffer Zone
A strip of land separating one type of land use from another, as a residential area is separated or shielded from a highway corridor. The buffer is intended to partially insulate the residential area from the noise and emissions emanating from the highway.

C

Capital Improvements
Items, including renovation and construction modifications, ordinarily treated as long-term investment (capitalized) because of their substantial value and life span.

Cease and Desist Order
A demand by a court or governmental agency that a firm or individual cease a specific, usually environmentally hazardous, activity.

CERCLA
See Comprehensive Environmental Response, Compensation and Liability Act.

CERCLIS
The Comprehensive Environmental Recovery, Compensation and Liabilities Information System; an EPA database which identifies hazardous waste sites that require investigation and possible remedial action to mitigate potential negative impacts on public health and the environment.

Cleanup
Actions taken to deal with a release or threatened release of hazardous substances that could affect the public health, welfare, and/or the environment. The term "cleanup" is often used generally to describe various implementation actions or phases of remedial responses, such as the remedial investigation and feasibility study (RI/FS).

Comprehensive Environmental Response, Compensation and Liability Act (CERCLA)
CERCLA, known as "Superfund," is a federal law passed in 1980 and modified in 1986 by the Superfund Amendments and Reauthorization Act (SARA). CERCLA established a tax on certain chemical feed stocks to be used to fund the cleanup of abandoned hazardous waste sites. The Superfund trust was designed to provide immediate remedial action to investigate and clean up abandoned, or uncontrolled, highly contaminated sites. The federal government can then seek to recover these costs through negotiation or legal action against the contributors or sources of pollution. Under the act, EPA can: pay for site cleanup when parties responsible for the contamination cannot be located or are unwilling or unable to perform the work; or take legal action to force parties responsible for the contamination to clean up the site or pay the federal government back for the costs of cleanup.

Contamination
The degradation of air, land, or water quality as a result of human activities.

Contract Lab
Laboratories under contract to EPA which analyze soil, water, and waste samples taken from areas at or near Superfund sites.

Corrective Action
Any action taken in order to come into compliance with any federal, state, or local statutory or regulatory requirement for the treatment, storage, or disposal of any hazardous waste.

Critical or Unique Habitat
The essential segment(s) of habitat that contains the unique combination of conditions (soils, vegetation, predator species, etc.) necessary for the continued survival of a endangered species.

D

Debris
1) A term applied to the loose material arising from the disintegration of rocks and vegetative material, transportable by streams, ice, or floods; 2) trash, ruins, and any accumulation of broken and detached matter, such as discarded household products and building materials.

Deed
A legal instrument in writing, duly executed and delivered, whereby the owner of real property (grantor) conveys to another (grantee) some right to title or interest in or to real estate.

E

Easement
A right or interest in the real property of another. The right to use another's land for a specific purpose, as a right of way.

Emergency Planning and Community Right-To-Know Act (EPCRA)
A 1986 Act requiring Governors to establish emergency response commissions, emergency planning districts, and local emergency planning committees that are to develop rules, procedures, and response plans in the event of local emergencies at facilities considered hazardous, by their nature or the characteristics of materials and chemicals used. EPCRA

establishes reporting requirements for owners and operators of facilities using hazardous chemicals; training programs; and public notification procedures.

Endangered Species
Those species in danger of extinction throughout all or a significant portion of their native habitat or range.

Environmental Data, Analyses
The sorting, selecting, comparing, decomposing, screening, simulating, weighing, rating, testing, computing, and segregating of environmental planning and management facts and information.

Environmental Database
A collection of information specific to an environmental operation, governmental agency, business, organization, pollution/hazard category, etc.

Environmental Inspector
A person who conducts field investigations to obtain environmental data for use by governmental, environmental and engineering and scientific disciplines to determine sources of pollution, extents of pollution, current environmental compliance status; recommends additional inspection or tests in the event of noncompliance findings. Usually conducts Phase I Environmental Inspections for a private firm or inspects for environmental compliance as a governmental staff member.

Environmental Mapping
A method of recording and displaying performance data in relation to an environmental setting. The planner notes activities and their frequencies at a site on facility maps, which, when completed, show operational behaviors. The data revealed from this technique includes location of storage, treatment, shipping, disposal and other activities such as frequency of the storage, treatment, shipping, disposal and other activities at varying locations (including vehicle, waste, and people movement, the relation of people to facility/project, and recurrent patterns of operations).

Environmental Protection Agency (EPA)
The federal agency established in 1970 to administer major laws to control and reduce contamination of air, water, and land systems.

Environmental Sampling
The process of scientific evaluation of site specimens to determine existence of, classifications of, quantifications of, extents of, and intensities of a contaminant; may be used to classify type where no contamination exists, as in a soil sample, to determine the bearing pressure of soils in an area to be developed. Tests are usually a combination of field work followed by an in-depth laboratory analysis.

EPA Hazardous Waste Number
The number assigned by EPA to each hazardous waste listed in Part 261, Subpart O, of 40 CFR and to each characteristic identified in Part 261, Subpart C, of 40 CFR.

EPA Identification Number
The number assigned by EPA to each generator, transporter, and treatment, storage, or disposal facility.

Erosion
Gradual deterioration of land formations by the action of wind, water, or by the construction/agricultural activities of humans. In the weathering process erosion is a natural

geological function, but more often than not it is poor land management that strips the land of its topsoil, leaving it vulnerable to erosion by wind and water.

Existing Use
A use existing at the time of enactment of a statute (ordinance) which does not comply with the legislation and which as a matter of due process of law, must be permitted to continue for some reasonable period of time.

F

Field Observation
Watching people's behavior, or physical changes occurring in a specific environmental setting. It is the most direct and reliable means for an environmental planner to get information about the way a facility and its personnel operate. Two data objectives of field observation are assessing activities and inventorying environmental settings and features. The focus is to measure and detail facility/project activity, settings, work group interactions, and relationships.

Field Photography
Photography, other than large-format photography, intended for the purpose of producing documentation, cataloging, and inventorying the natural and built environment through various annotated and dated photos. Field photography is used for: *recording* existing conditions to discern visible compliance status; *studying* selected site details, and sequential events such as the rate at which a leak spreads; and *analyzing* operational systems, as well as supplementing and verifying data.

Field Records
Notes of measurements taken, field photographs, and other recorded information intended for the purpose of producing documentation.

Flood Plain
A somewhat level part of land adjacent to a river channel, formed by sediments deposited by the river during repeated periods of flooding.

Freedom of Information Act (FOIA)
The federal act allowing public access to government documents and information.

Friable Asbestos-Containing Material
Material that can be easily crumbled by hand pressure.

G

Generator
Any owner or operator whose site activities, or processes, produces hazardous waste or whose actions first causes a hazardous waste to become subject to a regulation.

Grandfather Clause
Provision in a zoning ordinance which permits the continuance of property uses existing at the time the ordinance was passed, even though such uses are banned by the ordinance.

Greenbelt
Areas which may not be developed; generally serves as a buffer between pollution sources and concentrations of people.

Grid Pattern
A design layout used by urban planners based on spacing streets at regular intervals and intersecting them at right angles. The system is further enhanced by alphabetizing street names in one direction and numbering them in ascending order in the other.

Groundwater
Water that occupies pores and crevices in rock and soil, in a saturated zone or stratum beneath the surface of land or water. The upper limit of the groundwater is the water table.

Groundwater Flow
The direction and movement of water through openings in sediment and rock that occurs in the zone of saturation.

H

Habitat
The sum total of environmental conditions of a specific place occupied by an organism, population, or community; the dwelling place of a species with its particular environmental characteristics.

Hazardous Substance
Such chemicals, elements, compounds, substances and wastes, of any nature and composition, including toxic waste, as defined by any federal, state or local statute, regulation or ordinance for the protection of the environment, including, but not limited to, regulations promulgated pursuant to the Comprehensive Environmental Response, Compensation, and Liability Act of 1980, as amended. This typically includes hazardous substances and materials that are toxic, corrosive, ignitable, explosive, or chemically reactive, such as: petroleum and its by-products, asbestos, PCBs, radon, etc.

High-Concentration PCBs
An item, article, or material that contains 500 ppm or greater PCBs, or those materials that EPA requires to be assumed to contain 500 ppm or greater PCBs in the absence of testing.

Historic District
A geographically definable urban or rural area that possesses significant concentration, linkage, or continuity of historic sites, structures, or objects, unified historically or aesthetically by plan, physical developments, or similarity of use. A district also may be composed of individual elements by association or history. Historical units of the National Park System are ordinarily historic districts in themselves.

Historic Site
A distinguishable lot/property, with or without a structure; or an area of historic, prehistoric, or symbolic notoriety upon which an important historic or prehistoric event occurred, or which is directly connected with such events or persons, or which was the subject of an important historic or prehistoric activity.

Hydrogeologic
Those factors that deal with subsurface waters and related geologic aspects of surface waters.

Hydrology
The science dealing with the properties, movement, and effects of water on the earth's surface, in the soil and rocks below, and in the atmosphere.

I

Indigenous
Native or original to a specific area.

Indirect Source
A pollution source, such as a regional mall, not actually releasing the emission itself but attracting mobile emissions and resulting in an increase in pollutants in its immediate vicinity.

Industrial Waste
A term applied to the solid and liquid waste materials generated incidental to the various manufacturing processes employed at industrial plants and establishments, which require disposal. These wastes can be divided into five categories: 1) uncontaminated non-degradable factory refuse; 2) inert process waste; 3) flammable process waste; 4) either acid or caustic; or 5) indisputably toxic and hazardous waste. *Compare* Toxic Waste.

Interviews, Environmental
The most common, and most direct means of obtaining facility/project information. The purpose of conducting environmental interviews is to obtain specific environmental management and operations data on the facility and/or project, including management attitudes, operations activities, personnel tasks and competence, existing working/operating conditions, organization and perceptions of individuals participating as well as those of persons in the surrounding community.

J

Jurisdiction
The power of a governmental entity or court to review, permit, and determine compliance issues on specific types of project activities; the authority granted under a city's police powers to act over a particular activity.

L

Land Use
The deployment of real property for any legal use, by its owner of record, or a lessee.

Landfill
A disposal facility or part of a facility where waste is placed in or on land and which is not a land treatment facility, a surface impoundment, or an injection well.

Leak Detection System
A system capable of detecting the failure of either the primary or secondary containment structure or the presence of a release of hazardous waste or accumulated liquid in the secondary containment structure. A leak detection system must employ operational controls (daily visual inspections for releases into the secondary containment system of aboveground tanks) or consist of an interstitial monitoring device designed to detect continuously and

automatically the failure of the primary or secondary containment structure or the presence of a release of hazardous waste into the secondary containment structure.

Legal Description
A description recognized by law which is sufficient to locate and identify the property without oral testimony.

Liner
A continuous layer of natural or man-made materials, beneath or on the sides of a surface impoundment, landfill, or landfill cells, that restricts the downward or lateral escape of hazardous waste, hazardous waste constituents, or leachate.

M

Management of Migration
Actions that are taken to minimize and mitigate the migration of hazardous substances, pollutants, or contaminants and the effects of such migration. These actions may be appropriate when the hazardous substances are no longer at or near the area where they were originally located or situations where a source cannot be adequately identified or characterized. Measures may include, but are not limited to, provision of alternative water supplies, management of the contamination, or treatment of a drinking water aquifer.

Manifest
The shipping document EPA Form 8700-22 and, if necessary, Form 8700-22A, originated and signed by the generator in accordance with the instructions included in the appendix to Part 262 of the Resource Conservation and Recovery Act. The uniform shipping document is required by the EPA and establishes a tracking mechanism per RCRA. This tracking document follows the hazardous waste from point of generation to its final destination. Copies are maintained by the state where the wastes are generated, the destination state, the transporting company, and by the generator.

Material Safety Data Sheets (MSDS)
Documents required to accompany chemicals determined to be hazardous that contains information about the chemical's potential hazards and methods of controlling or preventing hazardous exposure.

Metes and Bounds
A method of describing or locating real property, metes are measures of length and bounds are boundaries. This description starts with a well-marked point of beginning and follows the boundaries of the land until it returns once more to the point of beginning.

Minimum Search Distance
The area for which records must be obtained and reviewed in accordance with the environmental database search of regulatory compliance records. The term may include areas outside the property and should be measured from the nearest property boundary. The Minimum Search Distance may be reduced by the Phase I inspector for all databases except the NPL Site List and the RCRA TSD List. Factors to consider in reducing the Minimum Search Distance include the density of, urban, rural or suburban of the setting in which the property is located; and the distance that the hazardous substance is likely to migrate based upon local geologic or hydrogeologic conditions.

Monitoring Wells
Special wells drilled at specific locations on or off a hazardous waste site where groundwater can be sampled at various depths and studied to determine such things as the direction in which the groundwater flows, and the types and amounts of contaminants present.

N

National Priorities List (NPL)
EPA's list of the most serious uncontrolled or abandoned hazardous waste sites identified for possible long-term remedial response using money from the "Superfund." The list is based primarily on the score a site receives on the Hazard Ranking System (HRS). EPA is required to update the NPL at least once a year.

National Register of Historic Places
A record of sites, structures, or objects of national, state or local significance. It was expanded by the 1966 Historic Preservation Act. To be eligible for inclusion on the National Register, properties can be publicly or privately owned but must meet the criteria found in 36 CFR 60 or 36 CFR 65. The program is administered by the National Park Service.

Noise Pollution
Excessive noise in the built environment.

Nonpoint Source
1) Generalized discharge of waste into the air, or water, whose specific source cannot be located; 2) a source discharging pollutants into the environment that is not a single point.

O

Onsite
1) On the same or geographically contiguous property. May be divided by public or private right-of-way, provided the entrance and exit between the properties are at a crossroads intersection, and access is by crossing, as opposed to going along, the right-of-way. 2) Noncontiguous properties owned by the same person but connected by a right-of-way controlled by that person and to which the public does not have access are also considered onsite property.

Open Dump
A solid waste land disposal site where wastes are disposed of in a manner that does not protect the environment, and where wastes are exposed to the elements.

Operations and Maintenance Plan (O&M Plan) for Asbestos-Containing Materials
Written instructions to building maintenance personnel on activities to be conducted at a site to avoid or minimize fiber release during activities affecting the ACM. At a minimum the O&M plan should define a periodic inspection program and provide precautions to be taken when working on or near ACM.

P

Particulates
A major category of airborne pollutants. Literally particles of dirt, dust, smoke, etc.

Parts Per Billion (ppb)
A unit of measure commonly used to express low concentrations of contaminants. For example: if one drop of chlorine (Cl) is mixed in an olympic size swimming pool, the water will contain approximately 1 ppb of Cl.

Parts Per Million (ppm)
A unit of measure commonly used to express low concentrations of contaminants (e.g., 1 ounce of Cl in 1 million ounces of water is 1 ppm).

Pathological Waste
Primarily hospital and laboratory wastes that contain pathogens, whose disposal must be carefully regulated due to the inherent risks to the public health and welfare.

PCB and PCBs
Any chemical substance that is limited to the biphenyl molecule that has been chlorinated to varying degrees or any combination of substances that contain any such substance (Polychlorinated Biphenyls).

Permeability
The ability of a porous medium to transmit fluids under a hydraulic gradient. The property or capacity of a porous rock, sediment, or soil for transmitting a fluid. A measure of relative ease of fluid flow under unequal pressure.

Phase I Environmental Site Assessment
An inspection process used to determine the environmental conditions at a specific site. At a minimum the Phase I process includes: a review of: historical site records; available files and databases maintained by regulatory agencies at all levels of government; an analysis of special resource issues; field inspections of the subject site and adjacent properties for potential hazardous waste problems; site photography; a preliminary comprehensive asbestos survey; a final report regarding any potential issues of concern, with appendices that include maps, aerial photos, a 50-year title chain, and pertinent agency findings on, and immediately adjacent to, the subject site. The Phase I inspection does not include sampling, monitoring, or other more detailed types of field/site investigations.

Phase II Environmental Site Assessment
An environmental process used to determine the extents of environmentally harmful chemical or solid wastes defined during a Phase I inspection, that pose a threat to public health. The Phase II process includes: 1) a review of the quantity and quality of Phase I data and site records; 2) a review of available files and databases maintained by regulatory agencies at all levels of government; 3) multiple field inspections to conduct site sampling and analysis of materials (or substances) on the subject site and adjacent properties; and 4) other monitoring as necessary, depending on specific site characteristics, including other more detailed types of field/site investigations.

Phase III Environmental Site Assessment
An environmental process used to mitigate the spread of environmentally harmful chemical or solid wastes sampled and analyzed during a Phase II Inspection, that pose a threat to public health. The Phase III process includes: 1) preparation of a remedial investigation plan to control, neutralize, and mitigate the threat; 2) site mobilization; 3) multiple field tasks to implement the mitigation alternatives outlined in the remediation plan, and alleviate the hazard by whatever means is required based on the specific site characteristics; and/or 4) other more complex types of field/site construction procedures to limit the spread of the hazard, pending further investigations.

Placards
DOT-required signs that are affixed to the front, rear, and sides of all vehicles transporting hazardous materials and/or wastes. These signs must meet DOT specifications for size, color, and location on the vehicle and must be used according to predesigned hazard classification criteria based on the weight or volume of the particular materials being transported.

Point Source
Any discernible, confined, and discrete conveyance, including, but not limited to, any pipe, ditch, channel, tunnel, conduit, well, discrete fissure, container, rolling stock, concentrated animal feeding operation, or vessel or other floating craft, from which pollutants are or may be discharged. This term does not include return flows from irrigated agriculture. Compare Nonpoint Source.

Potable Water
Fresh water that is safe for human use and consumption.

Potentially Responsible Party (PRP)
Any person, group of people, agencies, businesses, or combination of the same (such as owners, operators, generators, and transporters) who are potentially responsible for, or are contributing to, the contamination problems at a hazardous waste site. Whenever possible, EPA requires potentially responsible parties, through administrative and legal actions, to clean up the hazardous waste sites they have contaminated.

Prime Farmland
Land by its characteristics that is best suited for the production of food, feed, forage, fiber, etc.; land that has the soil quality, growing season, and moisture needed to economically produce sustained high yields of crops when treated and managed properly.

Public Trustee
The public official in each county, whose office has been created by statute, to whom title to real property is conveyed by Trust Deed for the use and benefit of the beneficiary, who usually is the lender.

Public Water Supply System
A system for providing piped water for human consumption that has at least 15 service connections and/or regularly serves at least 25 individuals on a daily or 60 days per annum basis. The term includes any collection, treatment, storage, and distribution facilities under control of the operator of such system and used primarily in connection with the system, and any collection or pretreatment storage facilities not under such control that are used primarily in connection with the system.

Q

Quadrangle
A tract of land in the U.S. government Survey System measuring 24 miles on each side of the square. Sometimes referred to as a "check."

Quarter Section
In U.S. and Canadian land surveying, a tract of land half a mile square containing 160 acres.

Quitclaim Deed
A deed in which the grantor warrants nothing. It conveys only the grantor's present interest in the real estate, if any.

R

RCRA
See Resource Conservation and Recovery Act of 1976.

RCRA Notifiers List
The EPA database which identifies companies which have notified the EPA (as required) that they generate, transport, treat, store, or dispose of hazardous wastes.

RCRIS
See Resource Conservation and Recovery Information System.

Real Property
Land; the surface of the earth and whatever is erected, growing upon, or affixed to the land; including that which is below it and the space above it. Synonymous with "land," "realty," and "real estate."

Records
Any report, document, writing, photograph, tape recording, or other electronic means of data collection and retention which pertains to defendant/respondent compliance with EPA, state and local govermental environmental rules and regulations.

Release
Any spilling, leaking, pumping, pouring, emitting, emptying, discharging, injecting, escaping, leaching, dumping, or disposing into the environment, including the abandonment or discarding of barrels, containers, and other closed receptacles containing any hazardous substance or pollutant or contaminant.

Remedy
1) The actual construction or implementation phase that follows the remedial design of the selected cleanup alternative at a site; 2) permanent remedy taken instead of, or in addition to, removal actions in the event of a release or threatened release of a hazardous substance into the environment; 3) to prevent or minimize the release of hazardous substances so that they do not migrate to cause substantial danger to present or future public health or welfare or the environment.

Representative
Any official of the owner or manager of a site, including receiverships, loan servicers, property managers, and any other designees.

Representative Sample
A sample of a universe or whole (e.g., waste pile, lagoon, groundwater) that can be expected to exhibit the average properties of the universe or whole.

Resource Conservation and Recovery Act of 1976 (RCRA)
RCRA is the federal statute enacted in 1976 regulating management of hazardous wastes to assure "cradle-to-grave" (or generation to disposal) responsibility and tracking. The statute was enacted as a result of the realization that the improper disposal of hazardous wastes posed a significant threat to human health and the environment. RCRA authorizes EPA to list waste materials as hazardous wastes and to develop record keeping, labeling, and handling requirements for hazardous waste. Most of the regulations developed under RCRA concern the control of hazardous waste generators, transporters, and treatment, storage, and disposal facilities (TSDPS).

Resource Conservation and Recovery Information System (RCRIS)
EPA's database of RCRA sites; includes small and large quantity hazardous waste generators, waste transporters, and waste treatment, storage, and disposal facilities.

Response Action or Response
Remove, removal, remedy, and remedial action involving either a short-term removal action or a long-term remedial response. All such terms (including the terms: removal and remedial action) include enforcement activities related thereto like: removing hazardous materials from a site to an EPA approved and licensed hazardous waste facility for treatment, containment, and/or destruction; containing the waste safely on-site to eliminate further spreading and problems; destroying or treating the waste on-site using incineration or other technologies; and/or identifying and removing the source of groundwater contaminants and halting further absorption of the contaminants.

Retention Pond
A man-made impoundment with a permanent pool of water that is used to reduce storm water flows at peak runoff.

Right-of-Way
An easement or right of passage over another's land; the strip of land used as roadbed by a railroad or used for a public purpose by other public utilities.

S

Sanitary Landfill
A facility for the disposal of solid waste which meets the criteria published under section 6944 of the Solid Waste Disposal Act.

SARA
An acronym for the Superfund Amendments and Reauthorization Act. *See* Comprehensive Environmental Response, Compensation and Liability Act (CERCLA).

SARA Title III Chemical Reporting
The EPA database from the Emergency Planning and Community Right-To-Know Act requirements which identifies a facility's chemical emissions, and the presence and amounts of hazardous chemicals maintained onsite.

Saturated Zone
That portion of the subsurface environment in which all voids are ideally filled with water under pressure greater than atmospheric; the zone in which the voids in the rock or soil are filled with water at a pressure greater than atmospheric. The water table is the top of the saturated zone in an unconfined aquifer.

Septic Tank
A watertight, covered receptacle designed to receive or process, through liquid separation or biological digestion, the sewage discharged from a building sewer. The effluent from such a receptacle is distributed for disposal through the soil, and settled solids and scum from the tank are pumped out periodically and hauled to a treatment facility.

Site Inspection (SI)
A technical phase that follows a preliminary assessment (PA) designed to collect more extensive information on a hazardous waste site. The information is used to score the site with the Hazard Ranking System to determine whether response action is needed or not.

Sketch Plan
A floor plan, generally not to exact scale, although often drawn from measurements, where the features are shown in proper relation and proportion to one another.

Sole-Source Aquifer
An aquifer that is the sole or principal source of drinking water, as established under Section 1424(e) of the Safe Drinking Water Act, and which if contaminated would create a significant hazard to public health.

Solid Waste
Wastes, including, but not limited to the following: *municipal wastes* - paper, metal, food, glass, yard wastes (grass clippings, tree trimmings, etc.), wood, plastics, cloth and rubber, and other inert material; *agricultural wastes* - animal wastes (manure), crop and orchard residues (straw, stubble, leaves, hulls, vines, tree limbs, etc.); *food processing wastes* - (animal parts, bones, fruit & vegetable pulp, seeds, skins, peelings, etc.), forest waste products - (sawdust, wood edgings, etc.); *mineral and fossil-fuel wastes* - copper, iron and steel, bituminous coal, phosphate rock, lead, zinc, etc.
As defined by RCRA, includes garbage, refuse, sludge from a waste treatment plant, water supply treatment plant or air pollution control facility; and other discarded material including solid, liquid, semi-solid, or contained gaseous materials resulting from industrial, commercial, mining, agricultural activities, and from community activities.

Source Reduction
The practice of reducing the amount of waste generated at the source of production. Examples include the redesign of processes to minimize waste volumes, the use of less hazardous or nonhazardous substitutes, and stack filtering and dewatering activities.

Special Resources
Site attribute which indicates a resource of natural, cultural, recreational, or scientific value of special significance, including, but not limited to: Archaeological resources that are listed or eligible for listing in the National Register of Historic Places under the National Historic Preservation Act; Undeveloped properties located within the boundaries of Sole Source Aquifers designated or proposed for designation, by the Environmental Protection Agency under the Safe Drinking Water Act; and Threatened or endangered species of plants and animals listed or proposed for listing, including their habitat, pursuant to the Endangered Species Act of 1973, as amended.

Spill
Intentional and unintentional spills, leaks, and other uncontrolled discharges; or 1) any quantity of PCBs running off or about to run off the external surface of the equipment or other PCB source as well as the contamination resulting from those releases. The concentration of the spill is determined by the PCB concentration in the material spilled as opposed to the concentration of the material onto which the PCBs were spilled; 2) if a spill of untested mineral oil occurs, the oil is presumed to contain PCBs greater than 50 ppm but less than 500 ppm; or 3) the uncontrolled leaking of any environmentally hazardous substance onto land or water.

State Historic Preservation Officer (SHPO)
An official within each state appointed by the governor to administer the State Historic Preservation Program. In addition, the SHPO has specific responsibilities relating to Federal undertakings that affect cultural resources within the state.

Suburbanization
The development, or urbanization, of formerly rural areas due to the out-migration of large numbers of people from urban to rural and suburban environments. *Compare* Urbanization.

Sump
Any pit or reservoir that meets the definition of tank including the trenches connected to it that seek to collect hazardous waste for transport to hazardous waste storage, treatment, or disposal facilities.

Superfund
Nickname for the Comprehensive Environmental Response, Compensation and Liability Act, also referred to as the Trust Fund. *See* CERCLA.

Surface Impoundment
A facility or part of a facility that is a natural topographic depression, man-made excavation, or dike area formed primarily of earthen materials (often lined with nonimpervious man-made materials), that is designed to hold an accumulation of liquid wastes or wastes containing free liquids, and that is not an injection well. Examples of surface impoundments include holding, storage, and settling ponds, aeration pits, and lagoons.

Surface Water
Bodies of water that are above ground, open to the atmosphere and subject to surface runoff (e.g., rivers, lakes, streams).

Surveys
Any systematic collection of primary data. There are varying types of survey techniques that can be applied in obtaining environmental information: the cross section, the longitudinal, and the contrasting sample survey are the most common socio-environmental types, while the characteristics of the physical environs are determined through engineering surveys.

T

Tank
A stationary device, primarily designed to contain/store accumulations of hazardous liquid wastes. Typically constructed of nonearthen materials (e.g., wood, concrete, steel, plastic) that provide impervious surfaces and structural support.

Tank System
A hazardous waste storage or treatment tank and its associated ancillary equipment and containment system.

Topography
The natural and artificial surface features of an area.

Topsoil
The surface layer of soil to a depth of about 1 foot.

Toxic Substances Control Act of 1976 (TSCA)
The federal statute designed to provide control over toxic chemicals before they reach commerce and industry. The Act requires prenotification to EPA of all new chemicals prior to manufacturing, including all existing data on toxicity and other characteristics, so that the EPA can regulate or prohibit the use of the chemical if it is found to pose an unreasonable

risk to human health and/or the environment. The Act requires the development of a comprehensive inventory of existing chemicals and substance testing.

Toxic Wastes
Solid, chemical, and liquid waste materials, generated incidental to manufacturing processes at a laboratory or industrial commercial site, which require disposal. These wastes can be categorized as either acid or caustic or as indisputably toxic and hazardous wastes. *Compare* Industrial Waste.

U

U.S. Fish and Wildlife Service (USFWS)
The branch of the U.S. Department of the Interior that is responsible for the protection of wildlife, wetlands habitats, and resource management.

U.S. Geological Survey (USGS)
An agency within the Department of the Interior responsible for conservation, geologic surveys, and mapping lands within the United States.

Underground Storage Tank (UST)
Any one of a combination of tanks, including pipes that connect them, used to contain an accumulation of regulated substances, and the volume of which is 10% or more beneath the surface of the ground.

Undeveloped
The term "undeveloped" means containing few manmade structures (less than one (1) aboveground, four-walled structure per five (5) acres when looking at the ratio of total acres of the site to total structures at the site) and having geomorphic and ecological processes that are not significantly impeded by any such structures or human activity.

Undeveloped Coastal Dunes and Beaches
Undeveloped coastal dunes and beaches that fall within the scope of the Coastal Zone Management Act.

Undeveloped Floodplains
Undeveloped floodplains as addressed in Executive Order 11988, "Floodplain Management," which is defined as the 100-year floodplain.

Urban Fringe
The transitional stretch of land undergoing a transmutation from rural to urban uses, usually promulgated by a major or regional type project; an area where population density is too high for rural, yet still too sparse for urban, infrastructure development and typical city services.

Urban Nodes
A planning concept whereby specific geographical areas are identified for development into centers of human work, shopping, and entertainment activities. This helps to consolidate infrastructure, transportation and other services.

Urbanization
The in-migration of people in large numbers from rural to urban environments. A characteristic of all industrialized countries. *Compare* Suburbanization.

W

Waste
Any substance, solid, liquid, or gas, that remains as a residue or is an incidental by-product of the use or manufacturing of a substance, which cannot be reused and/or disposed of in an environmentally safe manner.

Wastewater
The used water of a community or industrial operation. It may be a combination of the liquid- and water-carried wastes from residences, commercial buildings, industrial plants, and other sources, together with any groundwater, surface water and storm water that may exist.

Wastewater Lagoon
An impoundment area into which wastewater is discharged at a rate low enough to permit oxidation to occur without causing any substantial nuisances associated with odors and insects.

Water Table
The uppermost portion of groundwater whose level can vary based upon seasonal recharge and discharge rates. The water table can be measured by installing shallow wells extending a few feet into the zone of saturation and then measuring the water level in those wells.

Watershed
Area from which a stream or other body of water receives its makeup waters.

Well
Any shaft or pit dug or bored into the earth, generally of a cylindrical form, and often walled with bricks or tubing to prevent earth from caving in.

Wetlands
Low-lying lands which are near bodies of water, periodically covered by fresh, brackish, or salt water, and largely covered by vegetation. Wetlands usually include swamps, marshes, bogs, intermittent creeks and streams, and similar areas. These lands are subject to strict development restrictions.

Wild and Scenic River
A river and the adjacent area within the boundaries of a component of the Natural Wild and Scenic Rivers system.

Wilderness Area
Undeveloped federal land retaining its primeval character and influence, without permanent improvements or human habitation, that is protected and managed to preserve its natural conditions.

Wildlife Conservation
The practice of making decisions and implementing strategies to control wild animal populations and their environments, including influencing people to be cognizant of their impacts upon these resources.

Wildlife Habitat
Areas which provide food, shelter, and living environs for wildlife.

Z

Zoning
Land-use controls that regulate development uses, densities, height, bulk, lot proportions, etc. They are enacted by local governmental entities under the enabling legislation of the state. Zoning constitutes an exercise of police powers and must bear a reasonable relation to the preservation of public health, safety, and welfare.

Index

Index